Reliability Methods for Engineers

Reliability Methods for Engineers

K. S. Krishnamoorthi
Bradley University

ASQC Quality Press
Milwaukee, Wisconsin

Reliability Methods for Engineers
K. S. Krishnamoorthi

Library of Congress Cataloging-in-Publication Data
Krishnamoorthi, K. S.
Reliability methods for engineers / by K. S. Krishnamoorthi.
p. cm.
Includes bibliographical references.
ISBN 0–87389–161–9
1. Reliability (Engineering) I. Title.
TA169.K75 1992
620'.00452—dc20 92–184
CIP

10 9 8 7 6 5 4 3 2 1

ISBN 0–87389–161–9

Acquisitions Editor: Jeanine L. Lau
Production Editor: Mary Beth Nilles
Set in Times Roman by A-R Editions, Inc.
Cover design by Laura Bober
Printed and bound by BookCrafters

For a free copy of the ASQC Quality Press Publications Catalog,
including ASQC membership information call 800-952-6587.

Printed in the United States of America

 Printed on recycled paper

To Vijaya, Raja and Ram

CONTENTS

PREFACE

Reliability science deals with longevity of parts, products, and systems. When engineers are faced with questions such as how long a product will last, how the product life can be improved, what warranty can be given for a new product, when is the best time to overhaul a machine, or to replace a production tool, etc., they need reliability methods to find answers. All these questions relate to the random variable—product life, and statistical tools for data analysis and life prediction are necessary to answer such questions.

Reliability science has developed over the years into a sophisticated discipline and is well documented in some excellent books. Yet, knowledge of this science among working engineers outside of the electronics and defense industry is meager considering the need that exists. This is at least partly due to the lack of literature that extracts the essentials and presents them in a language the engineer can readily understand. This book attempts to fill that need.

The concepts are conveyed using just enough mathematics and statistics, while taking care not to oversimplify them. The mathematical insights (where given) will help in gaining confidence that comes out of knowing why a particular method or formula works the way it does. The insights also show the assumptions involved and the strength and limitations of the methods. This is meant to be a book of reliability by an engineer for engineers.

Whereas the variables in ordinary quality control inspection are likely to be normally distributed with its several well-known characteristics, the life variables are not so easily describable. Several distributions are needed to describe the time-variable and even for one product the failure time may follow different distributions during different periods of the product life. Furthermore, the nature of life data, coming from truncated tests and field experiments, tends to complicate the analysis. Statisticians have developed methods to handle many such complications.

A good understanding of the principles of probability and statistics is essential to gaining a good knowledge of reliability principles. Part I, therefore, is devoted to these basic principles, from which the reader can form a foundation for studying Parts II and III. Those who have prior education in probability and statistics may go directly to Part II after a cursory review of Part I.

Part II discusses the different distributions that have been found applicable as models for life-variables. Prediction methods based on these distributions are discussed in this section, as are methods of reliability apportionment.

Reliability tests are expensive because they require special fixtures and long testing periods, and destruction of some, if not all, sample units. This makes it necessary to properly plan the tests to acquire the most information from the least amount of testing. This is the subject of Part III. The MIL-HDBK-781D, a good source for reliability test plans, is covered in detail. This handbook requires quite a bit of statistical maturity on the user's part to utilize it correctly. The help provided in this section should aid the user in proper test plan selection.

Part III also has a brief description of nonstatistical methods used in accomplishing reliability goals. This portion of Part III also ties together various concepts discussed earlier showing how the different components fit together in the whole picture.

By no means is this a treatise on the state-of-the-art of reliability methods, but it can be considered a good starting point for those who want to learn these reliability methods and begin using them. Any comments or suggestions to improve the contents or method of presentation will be gratefully accepted.

I wish to thank Caterpillar Inc., Peoria, Il, for giving me permission to publish this book which is an outgrowth of a training manual I wrote for them. My special thanks are due to Robert B. Porter Jr of Caterpillar, Aurora, for his encouragement, criticisms, and support provided in writing the original manual.

K. S. K.
Peoria, Illinois
Nov. 1991

PART I

Fundamentals of Probability and Statistics

CHAPTER 1

Probability

. . . .

In order to understand the statistical methods in reliability, a good understanding of the fundamentals of probability and statistics is essential. An understanding of the meaning of the terms *probability*, *random variable, distribution, confidence interval, hypothesis testing*, etc., is necessary in order to fully appreciate the concepts and methods of reliability. Part I presents a quick overview of these fundamentals of probability and statistics before concepts of reliability are discussed in Part II.

Definition of Probability

A few preliminary definitions are necessary in order to define probability.

An *experiment* is a clearly defined procedure that results in observations. A single performance of an experiment is called a *trial*. Each trial of an experiment results in an *outcome* or *observation*.

The type of experiments dealt with herein are called *random experiments* in that the outcome in any one trial of such an experiment cannot be predicted with certainty. All possible outcomes of the experiment, however, will be known.

The set of all possible outcomes of a random experiment is called the *sample space* and is denoted by S.

Examples of experiments and sample spaces

Experiment 1: Throw a die and observe the number that shows.

S_1: {1, 2, 3, 4, 5, 6}

Experiment 2: Toss two coins and observe the faces on both.

S_2: {HH, HT, TH, TT}

Experiment 3: Toss two coins and count the number of heads.

S_3: {0, 1, 2}

Experiment 4: Take a sample of 10 pieces from a production line and observe the number of defectives.

S_4: {0, 1, 2,. . ., 10}

Each element of a sample space is called a *sample point*.

An *event* is a subset of the sample space such that all the elements in it satisfy a common rule. An event can be specified by the rule the elements satisfy or by enumerating all the elements in it.

Examples of events

EXPERIMENT	S	EXAMPLE EVENTS
1. Toss a die and observe the number	{1, 2, 3, 4, 5, 6}	*A*: {No. less than 4} *A*: {1,2,3}
2. Toss two coins and observe the faces	{HH, HT, TH, TT}	*B*: {At least one head} *B*: {HH, HT, TH}
3. Toss two coins and count the number of heads	{0, 1, 2}	*C*: {No head} *C*: {0}
4. Pick a sample of 10 items and count the number of defectives: D	{0, 1, 2, . . ., 10}	*E*: {"Accept" lot} ≡ {D < 3} *E*: {0, 1, 2}

Probability defined

Probability of an event is a number between 0 and 1, that indicates the likelihood of occurrence of the event when the associated experiment is performed.

Probability of an event that cannot occur = 0. Probability of an event that is certain to occur = 1.

We use the notation $P(A)$ to denote the probability of an event A. Thus, the definition of probability in terms of notations:

$$0 \leq P(A) \leq 1$$

$P(\Phi) = 0$ where Φ is the null event, the event that cannot occur

$P(S) = 1$

Next we will see how to compute the probability of an event.

Computing Probability of Events

Basically there are two methods of computing probability of an event.

1. Method of analysis of experiment
2. Method of relative frequency

There is a third method that involves subjectively assigning probability values to events based on one's experience with such events. Such probabilities are used in a branch of statistics known as Bayesian methods. We will not discuss Bayesian methods in this book.

Method of Analysis

Step 1: Formulate the S for the experiment.

Step 2: Assign weights to each of the elements in S such that the weights reflect the likelihood of occurrence of the elements when the experiment is performed, the weights are non-negative, and the weights add up to 1.

Such assigning of weights to the sample points in S requires a good understanding of the experiment. In some sample spaces the sample points all will have equal weights, in which case we call them equally likely elements. In other sample spaces the sample points will not be all equally likely.

Step 3: Calculate $P(A)$ as the sum of the weights of sample points in A.

Example Onc card is drawn from a deck. What is the probability that the card has a number (not a picture)?

Solution **Step 1**: Formulate S.

$S = \{$H2,. . .,H9,H10,HJ,HQ,HK,HA;
C2,. . .,C9,C10,CJ,CQ,CK,CA;
D2,. . .,D9,D10,DJ,DQ,DK,DA;
S2,. . .. S9,S10,SJ,SQ,SK,SA$\}$

Step 2: Assign weights to the elements.
Each one of these sample points is equally likely when a card is

drawn and, since there are 52 sample points in the sample space each has a weight of 1/52.

Step 3: Calculate $P(A)$.

The event A: {the card is a number} has 36 sample points in it and so, $P(A) = 36/52$

Example

A loaded die has even numbers twice as likely to show up as the odd numbers. What is the probability that the number that shows is less than 5 when such a die is thrown?

Solution

Step 1: $S = \{1, 2, 3, 4, 5, 6\}$

Step 2: From the fact the even numbers are twice as likely to occur, the weights for the six outcomes are:

$$\left\{\frac{1}{9}, \frac{2}{9}, \frac{1}{9}, \frac{2}{9}, \frac{1}{9}, \frac{2}{9}\right\}$$

Step 3: Event A: {# is less than 5} $\equiv \{1, 2, 3, 4\}$

$$P(A) = \frac{1}{9} + \frac{2}{9} + \frac{1}{9} + \frac{2}{9} = \frac{6}{9}$$

The Special Case

If the sample space of the experiment consists of elements that are all equally likely, then:

$$P(A) = \frac{\text{no. of elements in A}}{\text{no. of elements in S}} = \frac{N_A}{N_S}$$

For any experiment, the sample space can be written in a few different ways. Where possible, it is advisable to write the sample space consisting of equally likely outcomes so that the probability of events can be computed using this simple formula.

Method of Relative Frequency

This method is useful when the experimenter cannot assign weights to the different outcomes in S from prior knowledge of the experiment. Then the probability is determined through experimentation.

Step 1: Perform the experiment a certain number of times, say N, and find the number of times the event A occurs, say n.

Step 2: Calculate the relative frequency of the event A:

$F_A = n/N$

$P(A)$ is given by F_A provided N is sufficiently large;

that is: $P(A) = \lim_{N \to \infty} f_A$

Example A survey of 100 students at Bradley University engineering showed the following results.

	From Illinois Outside Chicago	From Chicago	From other areas
Boys	13	55	10
Girls	10	10	2

What is the probability a student picked at random in Bradley University engineering comes from Chicago? What is the probability the student is a girl and is from Illinois outside Chicago?

Solution Here, the experiment has been performed 100 times and the outcomes have been recorded. The relative frequency for each event can be taken as probability of that event since 100 trials can be considered sufficiently large. (The relative frequency will become a constant when the number of trials increase. When this occurs we say we have large enough trials.)

P (a student comes from Chicago) $= 65/100 = 0.65$

P (student is girl and is from Illinois outside Chicago) $= 10/100 = 0.1$

The above two methods would help in computing probability of simple events. We need to know some additional theorems for computing probability of more complex events which we will come across in quality control and reliability work.

Theorems on Probability

Theorem 1

If A and B are any two events in a sample space, then

$$P(A \text{ or } B) = P(A) + P(B) - P(A \text{ and } B)$$

Corr. If A and B are mutually exclusive, i.e., $P(A \text{ and } B) = 0$

$$P(A \text{ or } B) = P(A) + P(B)$$

Corr. If $A_1, A_2, \ldots, A_k$ are all mutually exclusive events, then

$$P(A_1 \text{ or } A_2 \text{ or } \ldots \text{ or } A_k) = P(A_1) + P(A_2) + \ldots + P(A_k)$$

When $A_1, A_2, \ldots, A_k$ are not mutually exclusive, the result is more complex and is not given here.

Example When a pair of dice is thrown, what is the probability the number 5 or 6 show?

Solution First, construct the sample space of the experiment.

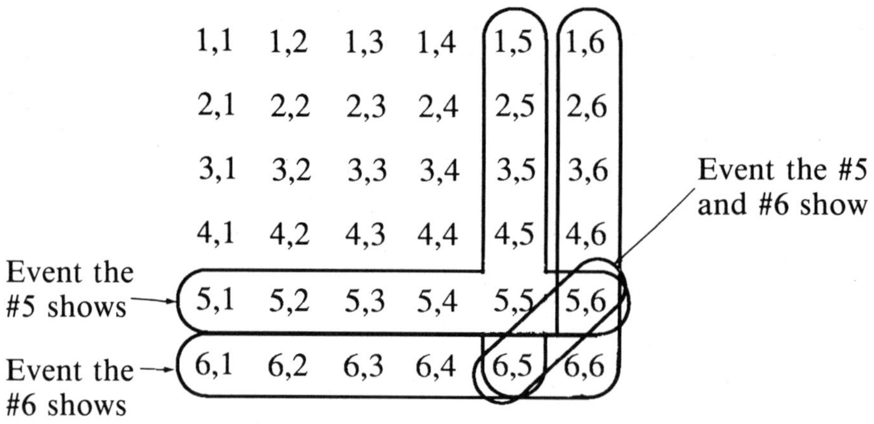

1,1 1,2 1,3 1,4 1,5 1,6

2,1 2,2 2,3 2,4 2,5 2,6

3,1 3,2 3,3 3,4 3,5 3,6

4,1 4,2 4,3 4,4 4,5 4,6

5,1 5,2 5,3 5,4 5,5 5,6

6,1 6,2 6,3 6,4 6,5 6,6

Event the #5 and #6 show

Event the #5 shows

Event the #6 shows

$$P(5 \text{ or } 6) = P(5) + P(6) - P(5 \text{ and } 6)$$
$$= 11/36 + 11/36 - 2/36 = 20/36 = 5/9$$

Example When a pair of dice is thrown what is the probability the total is less than 4 or one of the numbers is 4?

Solution Again, looking at the sample space,

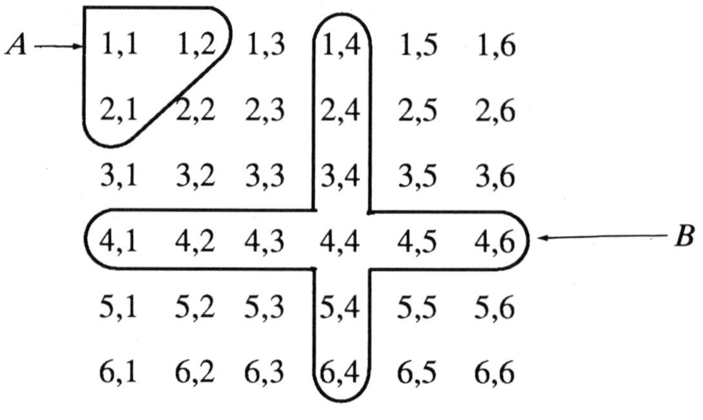

A → 1,1 1,2 1,3 1,4 1,5 1,6

2,1 2,2 2,3 2,4 2,5 2,6

3,1 3,2 3,3 3,4 3,5 3,6

4,1 4,2 4,3 4,4 4,5 4,6 ← B

5,1 5,2 5,3 5,4 5,5 5,6

6,1 6,2 6,3 6,4 6,5 6,6

Let A: {Total < 4}, then $P(A) = 3/36$
Let B: {one of the numbers is 4}, then $P(B) = 11/36$

A and B are mutually exclusive, i.e., when A occurs B cannot occur. Hence,

$$P(A \text{ or } B) = 3/36 + 11/36 = 14/36 = 7/18$$

Theorem 2

If A and A^c are complementary events, i.e., A and A^c are mutually exclusive and together make up the sample space.

$$P(A^c) = 1 - P(A)$$

Example When a coin is tossed six times, what is the probability that at least one head appears?

Solution $S = \{ \text{HHHHHH, HHHHHT,} \ldots \text{, TTTTTH, TTTTTT} \}$

There are $2^6 = 64$ equally likely outcomes in S

P (at least 1 head) $= 1 - P$ (no head) $= 1 - (1/64) = 63/64$

Independent Events

Two events in the same sample space are said to be independent if the occurrence of one event does not affect the chance of occurrence of the other. If independence between two events is obvious, the following theorem can be used to find the probability of joint or simultaneous occurrence of the events:

Theorem 3

If A and B are independent

$$P(A \text{ and } B) = P(A) \times P(B)$$

This theorem also can be used to determine if two events are independent when their independence is not obvious. The following two examples illustrate the use of this relationship.

Example An urn contains seven black balls and five white balls. If two balls are drawn with replacement (each ball is put back after observing its color), what is the probability both are black?

Solution Let B_1 be the event: the first ball is black. Then, $P(B_1) = 7/12$

Let B_2 be the event: the second ball is black. Then, $P(B_2) = 7/12$ because of replacement. It is easy to see B_1 and B_2 are independent. Hence,

$$P(B_1 \text{ and } B_2) = P(B_1) \times P(B_2) = \frac{49}{144}$$

Example Toss a pair of dice. Let E_1 be the event that the sum of the numbers is 6 and E_2 be the event when the sum is 7. Let F be the event when the first number is 3.

1. Are E_1 and F independent?
2. Are E_2 and F independent?

Solution

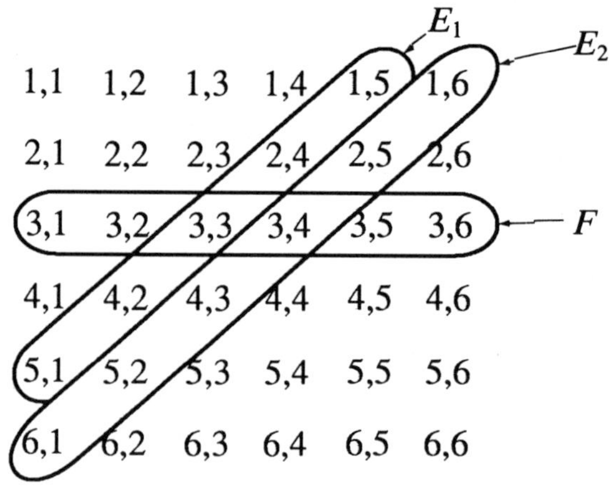

E_1 E_2

1,1	1,2	1,3	1,4	1,5	1,6
2,1	2,2	2,3	2,4	2,5	2,6
3,1	3,2	3,3	3,4	3,5	3,6
4,1	4,2	4,3	4,4	4,5	4,6
5,1	5,2	5,3	5,4	5,5	5,6
6,1	6,2	6,3	6,4	6,5	6,6

F

1. E_1 and F are independent if $P(E_1 \text{ and } F) = P(E_1) \times P(F)$

   $P(E_1 \text{ and } F) = 1/36$
   $P(E_1) = 5/36$
   $P(F) = 6/36$

   Since $P(E_1 \text{ and } F) \neq P(E_1) \times P(F)$, E_1 and F are not independent.

2. E_2 and F are independent if $P(E_2 \text{ and } F) = P(E_2) \times P(F)$

   $P(E_2 \text{ and } F) = 1/36$
   $P(E_2) = 6/36$
   $P(F) = 6/36$

   Since $P(E_2 \text{ and } F) = P(E_2) \times P(F)$, E_2 and F are independent.

This analysis shows that, whereas the total of the two numbers being 7 is independent of what happens to the first die, total being 6 is not independent of the first number because if the number 6 occurs in the first die the total cannot be equal to 6.

Statistical independence is an important concept and the above two examples are meant to illustrate it.

When there are two events that are not independent, their joint probability must be determined using conditional probabilities.

Conditional Probability

The conditional probability of an event A given that another event B (in the same sample space) has occurred is written as $P(A|B)$. It is obtained as the ratio of the sample points in both A and B, to those in B. This can be also seen to be, with reference to the sketch below, the proportion of sample points in B that are also in A. In terms of notations:

$$P(A|B) = \frac{P(A \cap B)}{P(B)}$$

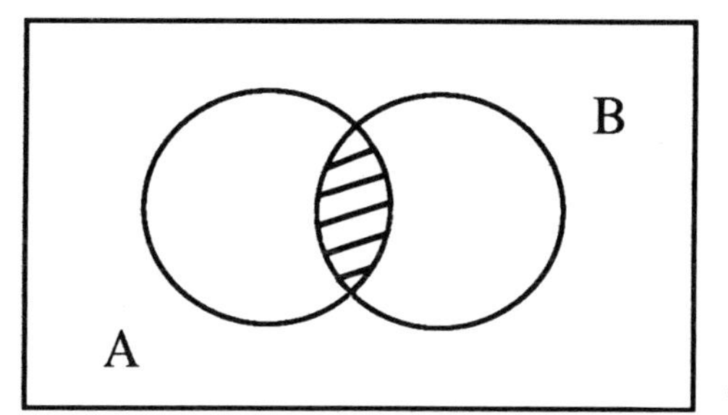

Example Find the probability that when two dice are thrown, the total equals 6 given one of the numbers is 3.

Solution

A

1,1	1,2	1,3	1,4	1,5	1,6	
2,1	2,2	2,3	2,4	2,5	2,6	
3,1	3,2	3,3	3,4	3,5	3,6	B
4,1	4,2	4,3	4,4	4,5	4,6	
5,1	5,2	5,3	5,4	5,5	5,6	
6,1	6,2	6,3	6,4	6,5	6,6	

Define A: total equals 6

B: one of the numbers is 3

We have to find $P(A \mid B)$

$$P(A|B) = \frac{P(A \cap B)}{P(B)} = \frac{1/36}{11/36} = \frac{1}{11}$$

Notice that $P(A \cap B)$ and $P(B)$ have been calculated with respect to the original sample space of the experiment.

Joint probability of non-independent events

Theorem 4

If two events A and B are not known to be independent,

$$P(A \cap B) = P(A|B)\, P(B)$$

This theorem follows from the definition of conditional probability $P(A|B)$.

Example When a pair of dice is thrown, what is the probability that the total is less than six and one of the numbers is 3 or 4?

Solution

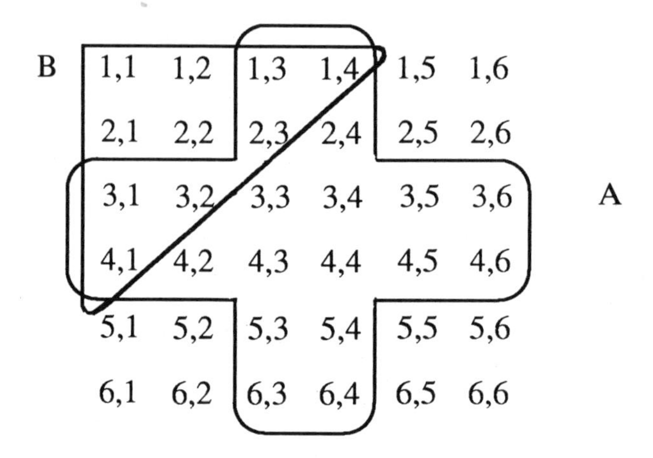

Let A: Event, one of the numbers is 3 or 4.
B: Event, the total is less than 6.

A and B are not independent.

$P(A \cap B) = P(A|B)\,P(B)$

$P(A|B) = 6/10, \qquad P(B) = 10/36$

$P(A \cap B) = (6/10)(10/36) = 6/36$

Although the answer could have been obtained directly from the sample space, the example illustrates the method of obtaining probability of joint occurrence of events that are not independent.

The value of obtaining probability of events conditioned on another event lies in the fact that, often times the probability of an event of interest is only known conditioned on the occurrence of another event whose probability is known. Sometimes the sample space is partitioned into sub-events that are mutually exclusive and together make the sample space. Then the probability of the event of interest conditioned on the partitions are used to obtain the probability of the event with respect to the total sample space. The *theorem on total probability* formalizes this procedure and the example that follows illustrates the concept.

Theorem on Total Probability

Let $B_1, B_2, \ldots\ldots B_k$ be partitions of a sample space S such that $B_1 \cup B_2 \cup \ldots\ldots \cup B_k = S$ and $(B_i \cap B_j) = \phi$ (null set) for any pair i and j (see diagram below). If A is an event of interest, then:

$$P(A) = P(A|B_1)\,P(B_1) + P(A|B_2)\,P(B_2) + \ldots\ldots + P(A|B_k)\,P(B_k)$$

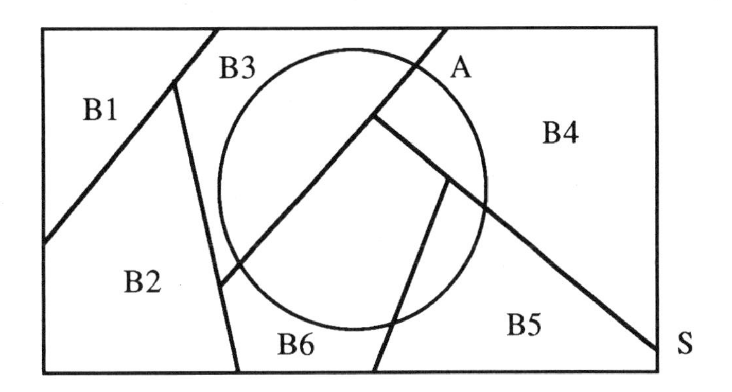

It is easy to see how the individual product terms on the right hand side of the above equation gives the joint probabilities of A with each of the partitions and, when they are added together they give the total, unconditioned probability of A.

Example

At Bradley University college of engineering the majors are distribusted as: 26% electrical, 25% mechanical, 18% civil, 12% industrial, and 19% manufacturing. Out of these, 5% of electrical, 10% of mechanical, 8% of civil, 45% of industrial, and 4% of manufacturing are women. If a student is picked at random in this college what is the probability that it is a women?

Solution

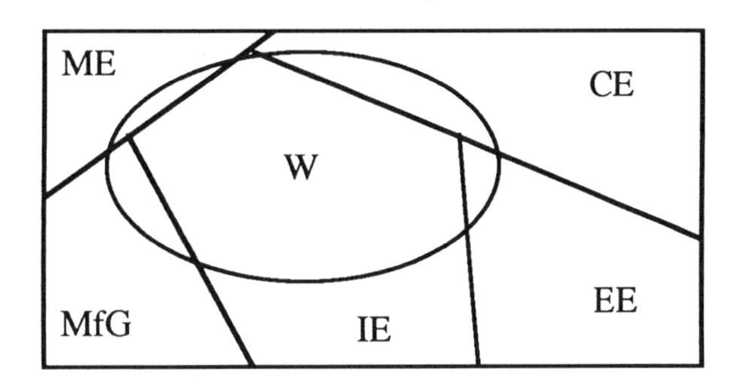

The student population is partitioned into the different engineering groups as shown in the diagram above.

When a student is picked at random $P(EE) = 0.26$, $P(ME) = 0.25$, $P(CE) = 0.18$, $P(IE) = 0.12$, $P(MfG) = 0.19$.

$P(\text{W} \mid \text{EE}) = 0.05$, $P(\text{W} \mid \text{ME}) = 0.10$, $P(\text{W} \mid \text{CE}) = 0.08$,
$P(\text{W} \mid \text{IE}) = 0.45$, $P(\text{W} \mid \text{MfG}) = 0.04$.

$$
\begin{aligned}
P(\mathrm{W}) &= P(\mathrm{W}|\mathrm{EE})\,P(\mathrm{EE}) + P(\mathrm{W}|\mathrm{ME})\,P(\mathrm{ME}) + P(\mathrm{W}|\mathrm{CE})\,P(\mathrm{CE}) \\
&\quad + P(\mathrm{W} \mid \mathrm{IE})\,P(\mathrm{IE}) + P(\mathrm{W} \mid \mathrm{MfG})\,P(\mathrm{MfG}) \\
&= (0.05)(0.26) + (0.10)(.25) + (0.08)(0.18) + (0.45)(0.12) \\
&\quad + (0.04)(0.19) \\
&= 0.114
\end{aligned}
$$

Counting Sample Points in a Sample Space

Sometimes finding the number of sample points in S presents a problem. The following approaches to counting provide help in such situations.

The Multiplication Rule

If an operation can be performed in n_1 ways and another operation can be performed in n_2 ways, then the two operations can be performed together in $n_1 \times n_2$ ways.

More generally: If operation 1 can be performed in n_1 ways, operation 2 in n_2 ways, . . . , operation k in n_k ways, then the k operations can be performed simultaneously in $n_1 \times n_2 \ldots \times n_k$ number of ways.

Example How many groups of two with one boy and one girl is possible from five boys and four girls?

Solution Consider this problem as filling two boxes, one to be filled with a boy and the other to be filled with a girl. Then, we want to find the number of ways in which both the boxes can be filled simultaneously.

There are five ways of filling the first box with a boy and four ways of filling the second with a girl.

[B] [G]

5 4

Both boxes can be filled together in $5 \times 4 = 20$ ways. There are thus 20 ways of choosing a group consisting of a boy and a girl, out of the five boys and four girls.

Example How many four-lettered words are possible from the five letters A, B, I, J, K if each letter can be used only once? How many of them will end with a vowel?

Solution Again, this problem can be looked upon as filling four boxes.

[1] [2] [3] [4]

$2 \times 3 \times 4 \times 5 = 120$

Starting with the fourth box, there are five possible ways of filling this box. After the fourth box is filled, there are four possible ways of filling the third box, etc. There are 120 ways in which the four boxes can be filled together. There are 120 four-lettered words possible from the five letters.

If the word must end with a vowel, the last box can be filled in only two ways. Thus, the number of ways in which the four boxes can be filled with this restriction:

[1] [2] [3] [4]
$2 \times 3 \times 4 \times 2 = 48$

Permutations

A permutation is an arrangement of all or part of a given set of objects. For example, the three objects a, b, and c can be permuted in six ways as:

abc, acb, bac, bca, cab, cba.

Similarly, there are six permutations of the three objects taken two at a time as follows:

ab, ba, ac, ca, bc, cb.

Theorem

The number of permutations of n objects taken r at a time:

$$_{n}P_{r} = \frac{n!}{(n - r)!}$$

When $r = n$, $_{n}P_{n} = n!$

Example How many starting lineups are possible with a team of 10 basketball players?

Solution

$$_{10}P_{5} = \frac{10!}{(10 - 5)!} = \frac{10!}{5!} = 10 \times 9 \times 8 \times 7 \times 6 = 30240$$

Example How many three-digit numbers can be formed from the digits 0, 1, 2, 3, 4, 5 if each digit is used only once?

Solution

$$_{6}P_{3} = \frac{6!}{(6 - 3)!} = \frac{6!}{3!} = 6 \times 5 \times 4 = 120$$

Combinations

A combination is a group of certain number of objects, taken from a given set of objects. Here, no attention is paid to relative location of objects in the group. For example, abc and acb are two different permutations, but they are one and the same combination.

Theorem

The number of combinations of n distinct objects taken r at a time is:

$$\binom{n}{r} = \frac{n!}{r!\,(n-r)!}, \qquad \text{for } r \leq n$$

Example How many different teams of five can be formed from a group of 10?

Solution

$$\binom{10}{5} = \frac{10!}{5!\,5!} = 252$$

Example How many different committees of three can be formed with two women and one man out of a group of four women and six men?

Solution Consider this problem as having to fill two boxes, the first box with two women and the second box with one man. The woman-box can be filled in $\binom{4}{2}$ ways and the man-box in $\binom{6}{1}$ ways.

The two boxes together can be filled in:

$$\overset{\text{W}}{\binom{4}{2}} \times \overset{\text{M}}{\binom{6}{1}} = \frac{4!}{2!\,2!} \times \frac{6!}{1!\,5!} = 6 \times 6 = 36$$

Example A box contains 10 pencils out of which four are good and six are bad.

a) How many samples of four can be drawn from the box?

$$\text{No. of samples of four} = \binom{10}{4} = \frac{10!}{4!\,6!} = 210$$

b) How many samples of four are possible with exactly one good?

$$\text{No. of samples with exactly one good} = \binom{6}{3} \times \binom{4}{1} = \frac{6!}{3!\,3!} \times 4 = 80$$

c) If a sample of four is drawn from the box, what is the probability that exactly one of them will be good?

$$\text{P(exactly one good)} = \frac{\binom{6}{3}\binom{4}{1}}{\binom{10}{4}} = \frac{80}{210} = 0.381$$

d) What is the probability that a sample of four will have no more than one good?

$$P(\text{no more than 1 good}) = P(0 \text{ good or } 1 \text{ good})$$
$$= P(0 \text{ good}) + P(1 \text{ good})$$
$$= \frac{\binom{6}{4}\binom{4}{0} + \binom{6}{3}\binom{4}{1}}{\binom{10}{4}}$$
$$= \frac{15 + 80}{210} = \frac{95}{210} = 0.452$$

EXERCISE

1.1 An experiment consists of throwing two dice and observing the numbers on both. An event E is said to occur if the difference between the two numbers is greater than 2. Find the elements in E.

1.2 An experiment consists of picking a student and measuring the circumference of head (X) and I.Q. (Y). Find the elements in the event $X < 24$ and $Y < 100$. (Use a graph.)

1.3 A die has three sides painted red, two sides painted yellow, and one side painted blue. What is the probability that when this die is thrown the face that comes up is not red?

1.4 An urn has three black balls and two white balls. If two balls are drawn from it without replacement, what is the probability that both are black?

1.5 When a pair of dice is thrown what is the probability that the number on the first die is 4 or the total is < 7?

1.6 When three dice are tossed what is the probability all three will not show the same number?

1.7 An unfair coin has $P(H) = 1/5$ and $P(T) = 4/5$. If this coin is tossed five times what is the probability that two out of the five tosses result in heads?

1.8 A manufacturer receives a certain piece part from four vendors in the following percentages: vendor A—28%, B—32%, C—18%, D—22%. Inspection of incoming parts reveal that 2% from vendor A, 1.5% from vendor B, 2.5% from vendor C and 1% from vendor D are defective. What percentage of the total supplies received by the manufacturer is defective?

1.9 **a)** In how many different ways can an eight-question true-false examination be answered without regard to being right or wrong?

b) If a student answers an examination as above at random, what is the probability that all eight answers will be correct?

1.10 In how many ways can four boys and three girls be seated in a row if the boys and girls must alternate?

1.11 a) How many three-digit numbers can be formed from the digits 0, 1, 2, 3, 4 if each digit can be used only once?

b) How many of them will be greater than 299?

1.12 An urn contains four green balls and three blue ones. What is the probability that two balls drawn from the urn without replacement will both be green?

1.13 Three defective items are known to be in a container containing 40 items. A sample of five items is selected at random without replacement.

a) What is the probability that the sample will contain no defectives?

b) What is the probability that the sample will contain exactly two defectives?

c) What is the probability that the number of defectives in the sample will be two or less?

CHAPTER 2

Random Variables and Distributions

A random variable is a variable that takes values at random, which means its next value cannot be predicted with certainty.

Examples of random variables:

1. The number that shows when a die is thrown.
2. The number of heads when a coin is tossed three times.
3. The number of bad widgets in a sample of 20.
4. The number of tosses of a coin needed to get five heads in a row.
5. Amount of ash in a pound of sugar.
6. Weight of a newborn baby in a hospital.
7. Circumference of head at temple of an adult.

There are two types of random variables: discrete and continuous.

1. A *discrete random variable* is one that takes a finite (or countably infinite) number of values:

The first four of the above examples are discrete variables. The fourth is a random variable that has countably infinite number of values.

2. A *continuous random variable* is one that takes infinite number of possible values. It takes values in an interval.

Examples 5, 6, and 7 above are continuous random variables.

There are several random variables in the natural and manmade world. Many times decisions must be made based on how the variables behave. Even though they seem to act in a haphazard manner, mathematicians studying their behavior have found that they follow certain rules of pattern. These patterns represented by formulas, pictures, tables, etc., used to portray the behavior of the random variables are called *probability distributions.*

Probability Distribution of a Discrete Random Variable

If X is a discrete random variable, a function $p(x)$ is defined as the *probability mass function* (pmf) of the random variable with the following properties:

1) $p(x) \geq 0$ for all x

2) $\Sigma_x p(x) = 1$

3) $p(x) = P(X = x)$

In other words, the $p(.)$ function is a non-negative function that gives the probability of the random variable taking different values. The sum of the function values over all possible values of X equals 1.0. Such a function is called the probability mass function of the discrete random variable.

Note that we use the capital letter X to denote the name of the random variable and the small case letter x to represent the values taken by X.

Example A random variable X denotes the number of heads when a coin is tossed three times. Find its probability mass function.

Solution Possible values for X: R_x: [0, 1, 2, 3]

The notation R_x denotes the *range space*, the set of all possible values of the random variable.

We find the probability of the variable taking the different values in R_x by making reference to the sample space of the experiment and identifying the equivalent events in the sample space that corresponds to the values in R_x.

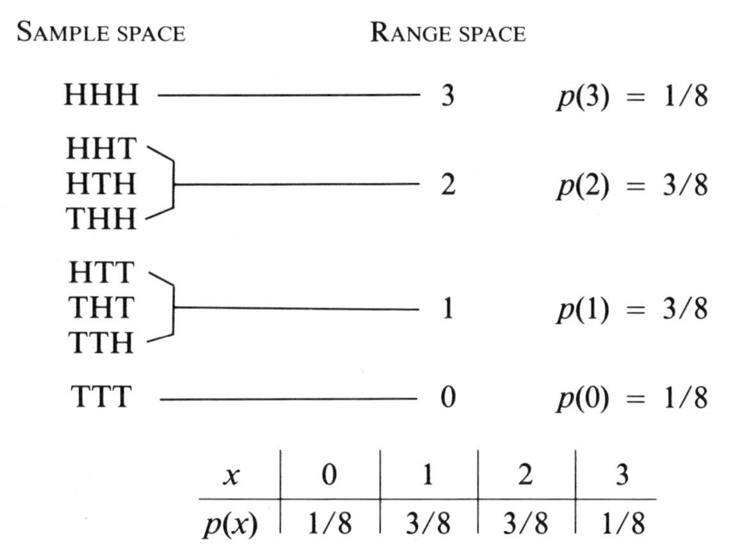

x	0	1	2	3
$p(x)$	1/8	3/8	3/8	1/8

This is the probability mass function of the random variable X. The function $p(x)$ satisfies all the required properties.

The above is a tabular representation of $p(x)$. It can be represented in a graph as shown in Figure 2.1

Or, it can be represented in a closed form as:

$$p(x) = \frac{\binom{3}{x}}{8}, \quad x = 0, 1, 2, 3$$

The closed form expression is concise and convenient. By any form, the $p(x)$ describes the behavior of the random variable and gives the chances of the random variable taking different values when it is observed next time.

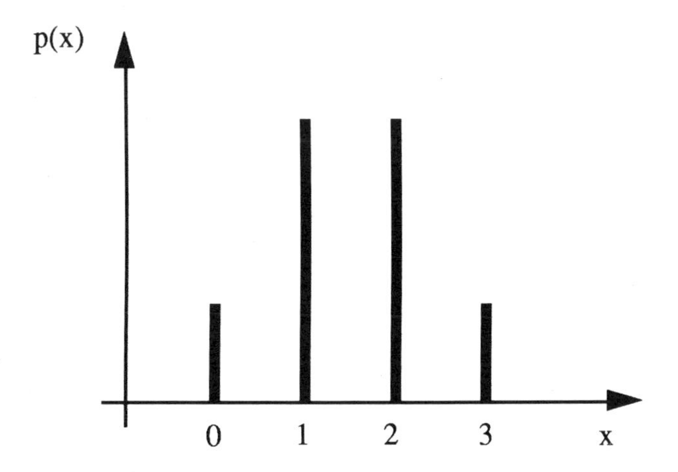

FIGURE 2.1 · *Graphical representation of a pmf*

Probability Distribution of a Continuous Random Variable

The method used to describe discrete random variables will not work in the case of continuous random variables because continuous random variables take an infinite number of possible values. The probability that a continuous random variable exactly equals one of the infinite possible values is *zero*. So, the probabilities of the random variable taking different values cannot be tabulated. However, we can consider the probability of the random variable taking values in an interval.

If X is a continuous random variable, a function $f(x)$ is defined with the following properties, and is called the probability density function (pdf) of X.

1. $f(x) \geq 0$ for all values of x
2. $\int_x f(x)dx = 1$
3. $P(a \leq X \leq b) = \int_a^b f(x)dx$

In other words, the function is positive valued over the range of possible values of the random variable. The area under the curve, which represents the function, over the possible values of the random variable is equal to 1.0. In addition, the area under the curve in any interval gives the probability of the random variable taking values in that interval.

Example A random variable X is known to have the pdf:

$$f(x) = \begin{cases} 0.01\,x & 0 \leq x \leq 10 \\ 0.01(20 - x), & 10 \leq x \leq 20 \\ 0, & \text{otherwise} \end{cases}$$

a) Verify if $f(x)$ is a valid pdf.

b) Find $P(5 \leq X \leq 10)$

Figure 2.2 shows the graph of the function

a) The function is positive valued. The area under the curve over all possible values of x, i.e., between 0 and 20 = area of the triangle = $(1/2)(20)(0.1) = 1.0$

b) $P(5 \leq X \leq 10)$ = Area under the curve between 5 and 10

$$= \frac{0.05 + 0.1}{2} \times 5 = \frac{0.75}{2} = 0.375$$

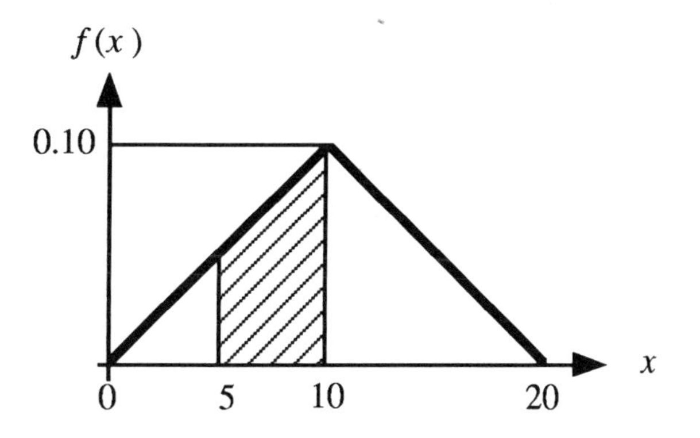

FIGURE 2.2 · *Graph of a pdf*

In this problem the area could be computed easily from basic geometry. If the graph of the function was not a simple geometric figure, integration must be used to compute the areas.

Characteristics of Distributions

A probability distribution completely describes the behavior of a random variable. However, for many distributions the description of the random variable can be accomplished using one, two, or three measures which would adequately represent those distributions. Below are some characterizing measures of distributions.

The Mean

If X is a discrete random variable with pmf $p(x)$, the mean:

$$\mu_x = \Sigma_x \, xp(x)$$

If X is a continuous random variable with pdf $f(x)$ then the mean:

$$\mu_x = \int_x x \, f(x)dx$$

The Variance

If X is a discrete random variable with pmf $p(x)$, the variance:

$$\sigma_x^2 = \Sigma_x \, (x - \mu_x)^2 \, p(x)$$

If X is a continuous random variable with pdf $f(x)$, the variance:

$$\sigma_x^2 = \int_x (x - \mu_x)^2 f(x)dx$$

Example If X represents the number of heads when a coin is tossed three times, find μ_x *and* $\sigma^2{}_x$.

Solution From an earlier example, the distribution of X is given by:

x	0	1	2	3
$p(x)$	$\frac{1}{8}$	$\frac{3}{8}$	$\frac{3}{8}$	$\frac{1}{8}$

$$\mu_x = 0 \times \frac{1}{8} + 1 \times \frac{3}{8} + 2 \times \frac{3}{8} + 3 \times \frac{1}{8} = \frac{12}{8} = \frac{3}{2}$$

$$\sigma_x^2 = (0 - \frac{3}{2})^2 \times \frac{1}{8} + (1 - \frac{3}{2})^2 \times \frac{3}{8} + (2 - \frac{3}{2})^2 \times \frac{3}{8} + (3 - \frac{3}{2})^2 \times \frac{1}{8}$$

$$= \frac{9}{4} \times \frac{1}{8} + \frac{1}{4} \times \frac{3}{8} + \frac{1}{4} \times \frac{3}{8} + \frac{9}{4} \times \frac{1}{8} = \frac{3}{4}$$

Example A random variable X has pdf $f(x) = \begin{cases} 2x & , 0 \leq x \leq 1 \\ 0 & \end{cases}$

Find μ_x and $\sigma^2{}_x$

Solution

$$\mu_x = \int_x x(2x)dx = 2\left(\frac{x^3}{3}\right)\Big|_0^1 = \frac{2}{3}$$

$$\sigma_x^2 = \int_x (x - \frac{2}{3})^2(2x)dx = \frac{1}{18}$$

From the definition of the mean μ_x and variance $\sigma^2{}_x$ we can see that the mean is the center of gravity of a distribution and the variance is the second moment of the distribution about the mean. The mean indicates where a distribution is located and the variance shows how dispersed the distribution is about the mean.

Standard deviation of a distribution is the (positive) square root of the variance. Thus, the standard deviation also is a measure of a distribution's variability. The standard deviation σ is generally used in industrial statistics as the measure of variability since its unit is the same as that of the variable.

There are several other measures used to characterize distributions (e.g., the standardized third moment and fourth moment, etc.), but the measures mean, variance, and standard deviation generally are adequate for distributions encountered in industrial statistics.

Some Important Probability Distributions

In the previous section we discussed the probability distributions as a means of describing behavior of random variables. These distributions are mathematical models that have to satisfy certain properties. A few such mathematical models have been found to be useful for describing random variables encountered in quality control and reliability work. The nature of these distributions and the context in which they are useful and some of their important properties are discussed below. The distributions that will be covered here are:

1. Binomial distribution
2. Poisson distribution
3. Normal distribution

Binomial and Poisson are discrete distributions while the Normal is a continuous distribution. The exponential and Weibull distributions, which are used in reliability work, will be discussed in Chapter 5.

Binomial Distribution

A random variable X is said to have the binomial distribution with parameters n and p if its probability mass function is given by:

$$p(x) = \binom{n}{x} p^x (1 - p)^{n-x}, \quad x = 0, 1, \ldots, n$$

Parameters of a distribution are the quantities that when known, completely describe the distribution. We will use the notation: $X \sim B(n, p)$ to indicate n and p are the parameters of the binomial distribution. This distribution is the model to use when describing random variables that represent the number of successes out of n independent trials, when each trial can result in success with probability p and failure with probability $(1 - p)$.

Examples of Binomial Random Variables

1. *X:* The number of heads when a fair coin is tossed 10 times:

$$X \sim B(10, 1/2)$$

2. *Y:* The number of baskets a ball player makes out of 12 free throws if his average is 0.4:

$$Y \sim B(12, 0.4)$$

3. *W:* The number of defectives in a sample of 20 taken from a (large) lot having 2 percent defectives:

$$W \sim B(20, 0.02)$$

It is necessary that the trials in the experiment be independent in order that the binomial model can be used to describe the random variable: number of successes. In the first example above, it is obvious that the trials are independent, because what happens in one toss does not affect what happens in other tosses. In the second example, it might be reasonable to make the assumption of independence among trials. In the third example, the independence requirement will not be met if sampling is done from a small lot, since what happens in picking one item will affect what happens in subsequent picks. As the lot size becomes larger compared to the size of the sample, however, this dependence becomes smaller, making the assumption of independence more valid. The use of the binomial distribution is illustrated in the following examples.

Example A sample of 12 bolts is picked from a production line and inspected. If it is known that the process produces 2 percent defectives, what is the probability the sample will have exactly one defective? What is the probability there will be no more than one defective?

Solution Let X represent the number of defectives out of 12.

$$X \sim B(12, 0.02)$$

$$p(x) = \binom{12}{x}(.02)^x(.98)^{12-x}, \text{ for } x = 0, 1, \ldots, 12.$$

$$P(X = 1) = p(1) = \binom{12}{1}(.02)^1(.98)^{11} = 12(.02)(.98)^{11} = 0.192$$

$$P(\text{no more than 1 def.}) = P(X \le 1) = p(0) + p(1)$$

$$= \binom{12}{0}(.02)^0(.98)^{12} + \binom{12}{1}(.02)^1(.98)^{11}$$

$$= 0.785 + 0.192 = 0.977$$

The Mean and Variance of a Binomial Variable

If $X \sim B(n,p)$, it can be shown using the definition for mean and variance that:

$$\mu_x = np$$

$$\sigma^2_x = np(1 - p)$$

Example If samples of 12 bolts are drawn repeatedly from a production line having 2 percent defectives, what will be the average number of defectives in the samples? What will be standard deviation of the number of defectives in the samples?

Solution

$$X \sim B(12, 0.02)$$

$$\mu_x = 12 \times 0.02 = 0.24$$

$$\sigma_x = \sqrt{np(1 - p)} = \sqrt{12 \times .02 \times .98} = 0.485$$

This means that if samples of 12 are repeatedly taken from the 2 percent lot over a long period of time, the number of defectives will average to 0.24; and the variability in those number of defectives will be given by $\sigma = 0.485$.

Poisson Distribution

A random variable X that has pmf given by:

$$p(x) = \frac{e^{-\lambda} \lambda^x}{x!}, \quad x = 0, 1, 2, \ldots$$

is called a Poisson random variable. The pmf is called the Poisson distribution.

We will use the notation: $X \sim Po(\lambda)$

This distribution has been found to be a good model to describe behavior of random variables such as:

1. Number of knots per sheet of plywood
2. Number of blemishes per shirt
3. Number of pinholes per square foot of galvanized sheet
4. Number of accidents per month in a factory, etc.

The Poisson distribution has one parameter and it equals the average value of the random variable. So, if we know the average value of the variable, we can find the probability of different events relating to the random variable.

Example A typist makes on the average three mistakes per page. What is the probability that the page she types for a typing test will have no more than one mistake?

Solution Let X be the number of mistakes per page

$$X \sim Po(3)$$

$$p(x) = \frac{e^{-3} 3^x}{x!}, \quad x = 0, 1, \ldots$$

$$\begin{aligned} P\text{ (no more than one defect)} &= P(X \leq 1) \\ &= p(0) + p(1) \\ &= e^{-3}\left[\frac{3^0}{0!} + \frac{3^1}{1!}\right] = e^{-3}[4] \\ &= 0.199 \end{aligned}$$

The mean and variance of a Poisson variable

If $X \sim Po(\lambda)$, it can be shown that:

$$\mu_x = \lambda \text{ and } \sigma^2_x = \lambda$$

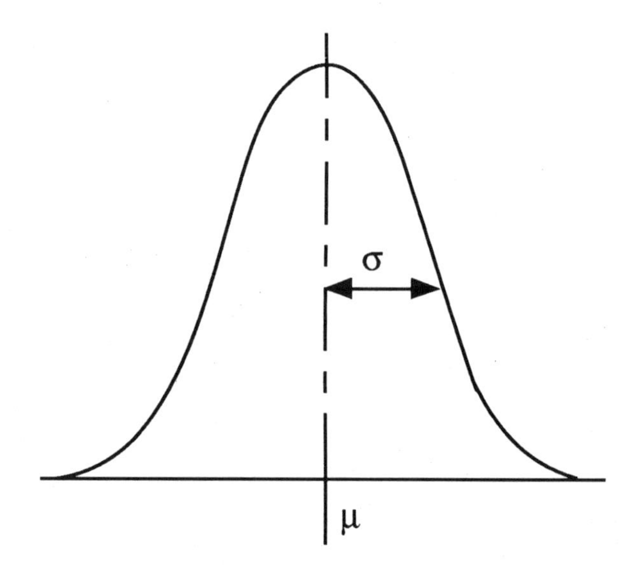

FIGURE 2.3 · *The graph of a normal pdf*

Poisson variable has the unique property that its mean and variance are equal.

Example If a typist makes on the average three mistakes per page, what would be the standard deviation of the mistakes per page in the long run by this typist?

Solution

$$X \sim Po\ (3)$$

$$\sigma_x = \sqrt{3} = 1.732$$

Normal Distribution

A random variable X is said to have the normal distribution with parameters μ and σ if its pdf is given by:

$$f(x) = \frac{1}{\sigma\sqrt{2\pi}} e^{-(1/2)[(x-\mu)/\sigma]^2}$$

The graph of the density function looks as shown in Figure 2.3.
The graph of the normal pdf known as the normal curve has certain special properties:

1. It is asymptotic with respect to the x-axis.
2. It is symmetric with respect to a vertical line at μ.
3. The maximum value of $f(x)$ occurs at μ.
4. The two points of inflexion occur at σ distance on each side of μ.

It can be shown that:

$$\int_x f(x)\, dx = 1 \qquad \text{[Area under the curve = 1]}$$

$$\int_x x\, f(x)\, dx = \mu \qquad \text{[Mean of the random variable = } \mu\text{]}$$

$$\int_x (x - \mu)^2 f(x)\, dx = \sigma^2 \quad \text{[Variance of the random variable = } \sigma^2\text{]}$$

We use the notation: $X \sim N(\mu, \sigma^2)$ to denote that the random variable X has normal distribution with parameters μ and σ^2. By specifying values for μ and σ we can completely describe the distribution of a normal random variable.

Normally distributed random variables occur frequently when dealing with industrial data. Most *measurements* are normally distributed. Length of bolts, diameter of bores, strength of wire, percent impurity, etc.; are examples of measurements that can be expected to be normally distributed.

Often a variable will be known to have the normal distribution with a known average and known standard deviation; the probability the variable lies within a given specification must be found. As an example, let $X \sim N(20, 9)$, and it is needed to find $P(10 \leq X \leq 15)$.

[A little reflection would help in accepting the interpretation of the probability of the variable within an interval as the proportion of values in the population lying in that interval.]

This probability is given by the area between 10 and 15 under the curve defined by the function:

$$f(x) = \frac{1}{3\sqrt{2\pi}}\, e^{-\frac{1}{2}\left(\frac{x-20}{3}\right)^2}$$

It is not easy to find this area using calculus. A different approach is adopted.

A random variable that is normally distributed with mean equal to 0 and variance equal to 1 is called a *standard normal variable*. Such a variable is denoted by Z. So: $Z \sim N(0, 1)$

Areas under the standard normal curve from $-\infty$ up to many z values have been tabulated as in Table A.1 in the appendix called the Normal Table.

There is a relationship which enables conversion of a problem relating to any normal distribution into one of standard normal distribution. First let's see how to use the Normal Table.

Examples in using the Normal Table:

If $Z \sim N(0, 1)$

a) Find $P(Z \leq 2.62)$

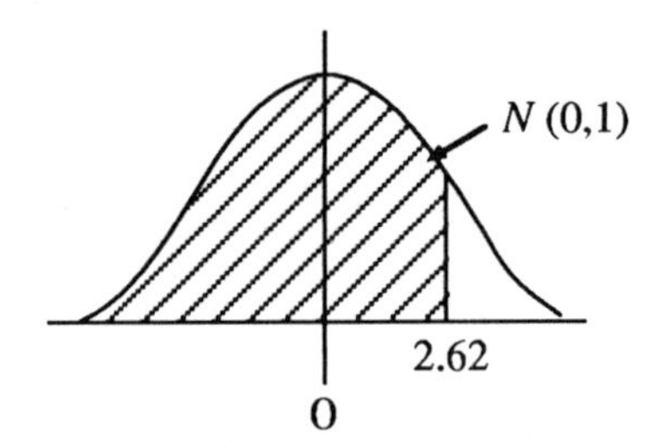

A sketch always helps in problems relating to normal distribution. With reference to the sketch above we need the area under the curve shown hatched. These areas are given in the normal table. From table, $P(Z \leq 2.62) = 0.9956$

b) Find $P(Z \leq -1.45)$

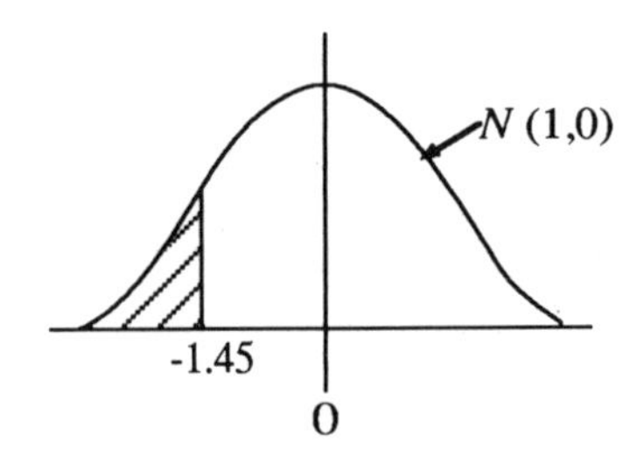

From the table, $P(Z \leq -1.45) = 0.0735$

c) Find $P(Z > 1.45)$

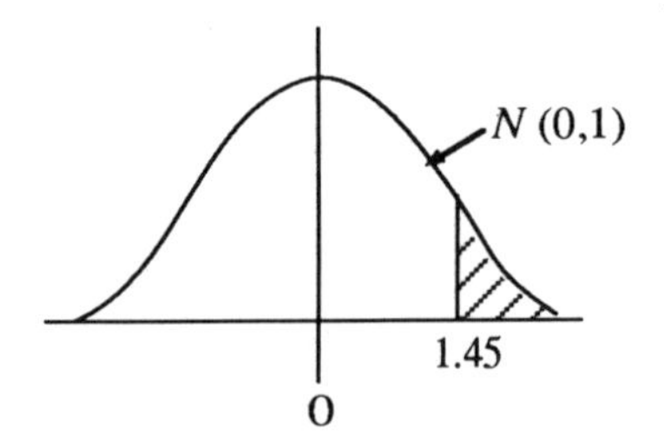

$P(Z > 1.45) = 1 - P(Z \leq 1.45) = 1 - 0.9265 = 0.0735$

Note $P(Z > 1.45) = P(Z < -1.45)$. This should be expected because of the symmetry of the normal curve.

d) Find $P(-1.5 \leq Z \leq 2.5)$

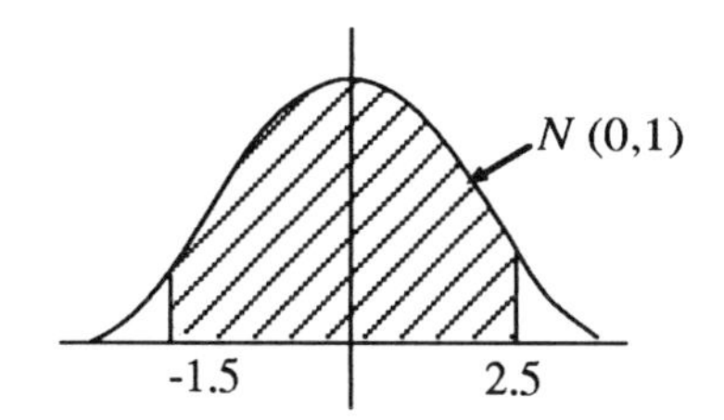

$$P(-1.5 \leq Z \leq 2.5) = P(Z \leq 2.5) - P(Z \leq -1.5) = 0.9938 - 0.0668$$
$$= 0.9270$$

e) Find t such that $P(Z < t) = 0.0281$

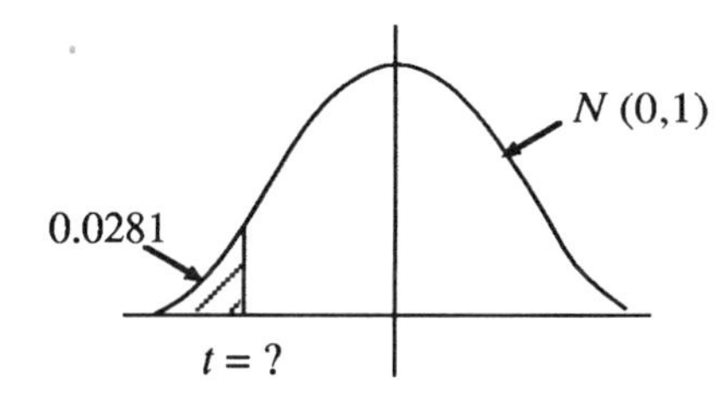

From the table, $t = -1.91$

f) Find s such that $P(Z > s) = 0.0771$

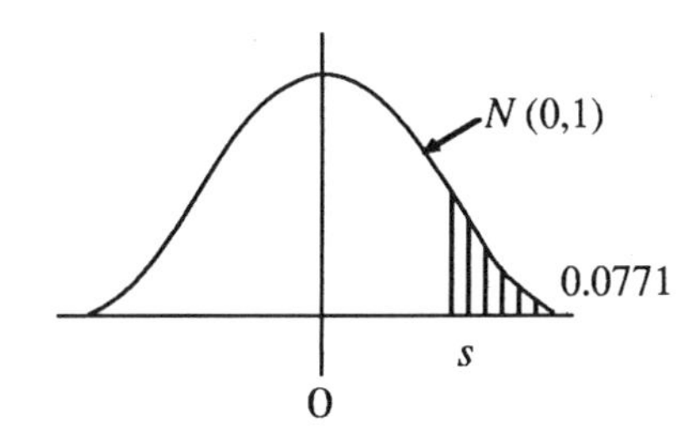

i.e., find $-s$ such that $P(Z \leq -s) = 0.0771$
$-s = -1.425, \qquad s = 1.425$

g) Find k such that $P(-k \leq Z \leq k) = 0.9973$

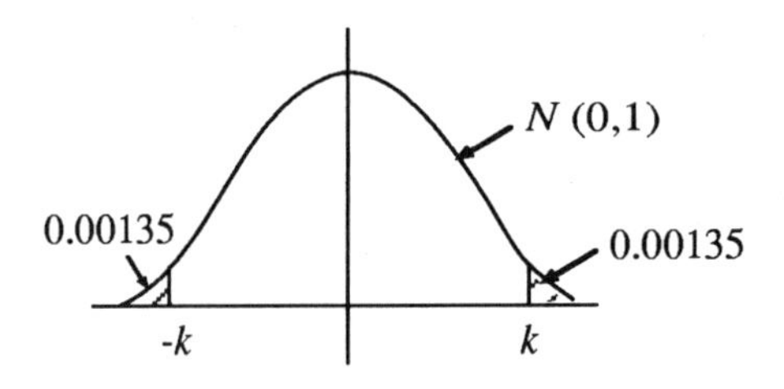

Because of symmetry $P(Z \leq -k) = P(Z > k) = 0.00135$
From the table, $k = 3$

The following theorem provides the link between any normal distribution and the standard normal distribution.

Theorem

If $X \sim N(\mu, \sigma^2)$; then

$$\frac{X - \mu}{\sigma} \sim N(0, 1)$$

This says, if there is a random variable X which is normally distributed, then a function of it, $\frac{X - \mu}{\sigma}$ (a function of a random variable is also a random variable), has the standard normal distribution. The following examples show how this relationship can help in finding areas under any normal curve.

A random variable $X \sim N(2.0, 0.0025)$

i.e., $\mu_X = 2.0 \quad \sigma^2_X = 0.0025 \qquad \sigma_X = 0.05$

Find (1) $P(X \leq 1.87)$

$$P(X \leq 1.87) = P(\frac{X - 2.0}{0.05} \leq \frac{1.87 - 2.0}{0.05})$$

$$P(Z \leq -2.6) = 0.0047$$

(2) $P(X > 2.2)$

$$P(X > 2.2) = P(Z > \frac{2.2 - 2.0}{0.05})$$

$$= P(Z > 4.0) = 0$$

(3) $P(1.9 \leq X \leq 2.1)$

$$P(1.9 \leq X \leq 2.1) = P(\frac{1.9 - 2.0}{0.05} \leq Z \leq \frac{2.1 - 2.0}{0.05})$$

$$= P(-2 \leq Z \leq 2) = P(Z \leq 2) - P(Z \leq -2)$$

$$= 0.9772 - 0.0228 \neq 0.9544$$

(4) Find t such that $P(X > t) = 0.05$

$$P(Z > \frac{t - 2}{0.05}) = 0.05$$

$$\Rightarrow \frac{t - 2}{0.05} = 1.645 \Rightarrow t = 2.08225$$

(5) Find k such that $P(\mu - k\sigma \leq X \leq \mu + \text{k}\sigma) = 0.9973$

$$\Rightarrow P(\frac{\mu - k\sigma - \mu}{\sigma} \leq Z \leq \frac{\mu + k\sigma - \mu}{\sigma}) = 0.9973$$

$$\Rightarrow P(-k \leq Z \leq k) = 0.9973$$

$$\Rightarrow k = 3$$

This means, for all normal distribution, 99.73 percent of the population fall within 3σ distance on either side of the mean. Similarly it can be shown that for all normal distribution, 95.44 percent of the population fall within 2σ distance on either side of the mean and 68.26 percent fall within σ distance on either side of the mean.

Application of Normal Distribution:

Example

1. Diameters of bolts in mass production are known to be normally distributed with mean = 0.25 inch and standard deviation 0.01 inch. Bolt specs call for 0.24 ± 0.02 inch.

a) What proportion of the bolts are outside spec?

b) If the process mean is moved to coincide with the center of spec, what proportion will be defective?

Solution

Let D be the diameter of bolts. Thus, $D \sim N(0.25, 0.01^2)$.
[It is a convention that the variance, not the standard deviation that is used as the parameter in this notation.]

a) We need $P(D < 0.22) + P(D > 0.26)$

$$= P(Z < \frac{0.22 - 0.25}{0.01}) + P(Z > \frac{0.26 - 0.25}{0.01})$$

$$= P(Z < -3) + P(Z > 1) = 0.00135 + 0.1587 = 0.16$$

i.e., 16 percent of the bolts are outside specification.

b) When process mean coincides with spec mean:

$$P(D < 0.22) + P(D > 0.26)$$

$$= P(Z < \frac{0.22 - 0.24}{0.01}) + P(Z > \frac{0.26 - 0.24}{0.01})$$

$$= 2 \times 0.0228 = 0.0456$$

i.e., 4.56% will be outside specification.

This brings home the point that, in general, centering a process will lead to considerable reduction in out-of-spec products.

Example

2. A battery manufacturer will replace any battery that dies before three years. It is known that the life of the batteries is normally distributed with mean = 4 years and standard deviation = 0.45 year.

a) What percent of the batteries will need replacement?

In battery manufacturing it is difficult to increase the mean life, but the variability can be changed with some effort.

b) What should be the standard deviation of battery life if no more than 0.02 percent of the batteries should require replacement?

Solution

Let X be the random variable that denotes the life of the batteries in years. $X \sim N(4, 0.45^2)$

a) $P(X < 3) = P(Z < \frac{3 - 4}{0.45}) = P(Z < -2.22) = 0.0132$

i.e., 1.32 percent will need replacement.

b) Let σ' be the new standard deviation

Find σ' such that $P(X < 3) = 0.0002$

$$\Rightarrow P(Z < \frac{3 - 4}{\sigma'}) = 0.0002$$

$$\Rightarrow \frac{3 - 4}{\sigma'} = -3.49$$

$$\Rightarrow \sigma' = \frac{1}{3.49} = 0.2865$$

36 percent reduction in variability is needed.

Distribution of the Sample Average

The sample average is a random variable. Each time a sample is taken from a population, the average assumes a different value. This variability is described by its distribution which has a mean and a variance. The following theorem shows the relationship between the distribution of a population and the distribution of the average of samples taken from it.

Theorem:

$$\text{If } X \sim N(\mu, \sigma^2)$$

$$\overline{X}_n \sim N(\mu, \frac{\sigma^2}{n}).$$

This theorem says: If we take samples of size n from a population that is normally distributed and compute averages, these averages also will have normal distribution with mean equal to the mean of the parent population and variance reduced depending on the sample size.

The above theorem gives the distribution of $\overline{X}$ for normal populations only. The next theorem gives a more general result that is applicable to all populations.

Example

Diameters of bolts are known to be normally distributed with mean = 2.0 inches and standard deviation = 0.15 inch. If a sample of nine bolts is drawn, what is the probability the average of these bolt diameters will not be greater than 2.1 inches?

Solution

Diameter of bolts: $X \sim N(2.0, 0.15^2)$

$$\overline{X}_9 \sim N(2.0, 0.05^2)$$

$$\mu_{\bar{x}} = 2.0$$

$$\sigma_{\bar{x}}^2 = \frac{0.15^2}{9} = 0.05^2$$

$$\sigma_{\bar{x}} = 0.05$$

We need $P(\overline{X} \leq 2.1) = P(Z \leq \frac{2.1 - 2.0}{0.05}) = P(Z \leq 2) = 0.9772$

Example

If samples of size four are taken from a population that is $N(2.0, 0.15^2)$ where should the two limits be set such that the probability the $\overline{X}$ values fall outside the limits does not exceed 0.0027? The two limits are set equidistant on either side of the mean.

Solution

$$\overline{X} \sim N(2.0, 0.15^2)$$

$$\mu_{\bar{x}} = 2.0$$

$$\sigma_{\bar{x}} = \sqrt{(0.15^2/4)} = 0.15/2 = 0.075$$

In order to include 0.9973 probability the limits must be set at 3σ distance where the σ is $\sigma_{\bar{X}} = 0.075$. Therefore the limits must be at $2.0 \pm 3 \times 0.075$, i.e., at 1.775 and 2.225. This, incidentally, is the principle behind determining the limits for the 3-sigma control charts used in process control.

The Central Limit Theorem

Let a population have any distribution. If samples of size n are drawn from this population, the $\bar{X}$s from such samples will have a distribution that approaches normal distribution if n is sufficiently large. In mathematical notations:

$$\text{If } X \sim f(x) \text{ with mean } = \mu, \text{ variance } = \sigma^2$$

$$\text{Then, } \bar{X}_n \xrightarrow[n \to \infty]{} N(\mu, \frac{\sigma^2}{n})$$

We can restate the theorem as: No matter what the process distribution is, the sample averages will be approximately normally distributed for large samples. It is known that this happens even for sample sizes as small as four. This is what makes many of the statistical methods used in inference robust with respect to process distribution. That is, although the inference procedures (discussed in the next chapter) based on $\bar{X}$ make the assumption that the population is normally distributed, some deviation from normality in the population does not seriously affect the inferences obtained from them.

EXERCISE

2.1 If Y denotes the sum of the two numbers when a pair of dice is thrown, find its pmf.

2.2 An urn contains two white balls and two black balls. If X denotes the number of black balls in a sample of three drawn from it, find the pmf of X.

2.3 A random variable X has pdf:

$$f(x) = \begin{cases} \dfrac{2(1 + x)}{27} & , 2 \leq x \leq 5 \\ 0 & , \text{otherwise} \end{cases}$$

find $P(X \leq 4)$

2.4 If Y denotes the sum of the two numbers when a pair of dice is thrown, find μ_Y and $\sigma^2{}_Y$.

2.5 An urn contains two white balls and two black balls. If X denotes the number of black balls in a sample of three balls drawn from it find μ_X and $\sigma^2{}_X$.

2.6 A plane has four engines. Each engine has probability 0.3 of failing during a flight. What is the probability that no more than two engines will fail during a flight? Assume the engines are independent.

2.7 A chemical plant experiences on the average three accidents per month. What is the probability that there will be no more than two accidents next month?

2.8 Let $X \sim N(10, 25)$. Find $P(X \leq 15)$, $P(X \geq 12)$, $P(9 \leq X \leq 20)$

2.9 Let $X \sim N(5, 4)$. Find b such that $P(X > b) = .20$

2.10 Let $X \sim N(5,9)$. Find the values a and b such that $P(a < X < b) = 0.874$ where the interval a to b is symmetric about the mean.

2.11 The life of a particular type of battery is normally distributed with mean 600 days and standard deviation 49 days. What fraction of these batteries would be expected to survive beyond 580 days?

2.12 Automatic fillers are used to fill 10-ounce cans of deodorants. The process standard deviation is 0.20 ounce. To ensure that every can meets this 10 ounce minimum, the company has set a target value for the process at 11.0 ounces.

(a) At the process average of 11 ounces, what percent of the cans will have less than 10.0 ounces of product? Assume weight of contents is normally distributed.

(b) Assuming that $\pm 3.5\sigma$ tolerance limits on the process cover virtually all of the filled cans, what is the minimum value to which the process average may be lowered in order to ensure that virtually no can is filled with less than 10 ounce?

CHAPTER 3

Statistical Methods for Inference

These methods are used to draw inference or conclusions about parameters of populations, from observations obtained from samples. The populations are assumed to have the normal distribution and hence the two parameters μ and σ^2 are of interest.

Figure 3.1 depicts the broad category of methods available for statistical inference. The objective of this chapter is to give a basic understanding of the concepts involved in the two main statistical methods estimation and hypothesis testing, because of their widespread use in reliability studies.

Definitions

A *statistic* is a function of observations from a sample. Sample average, $\overline{X}$, range R, standard deviation S, and the largest value $X_{\max}$, etc., are examples of statistics. A statistic is a random variable.

A statistic that is used to estimate a parameter is called an *estimator* for the parameter. For example, $\overline{X}$ or M (median) can be used as estima-

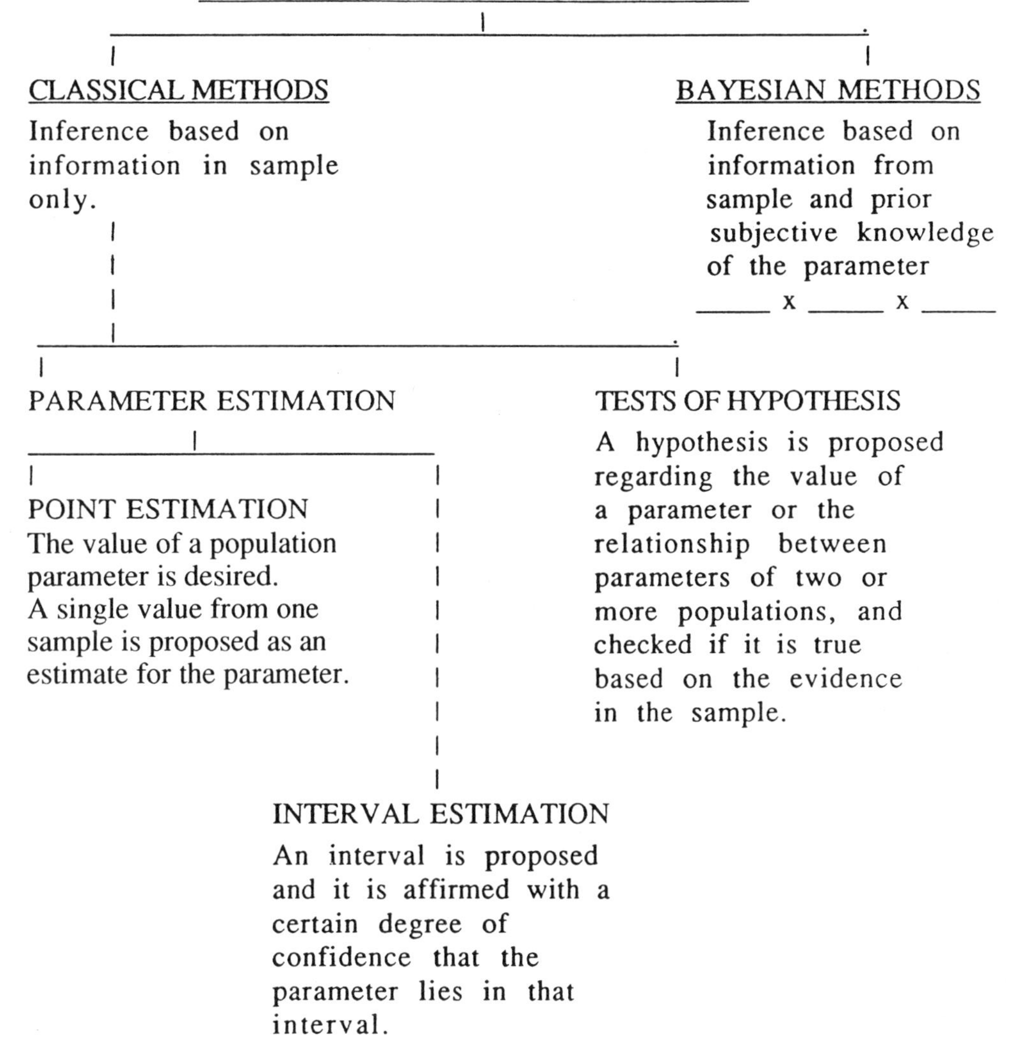
STATISTICAL METHODS FOR INFERENCE
CLASSICAL METHODS
Inference based on information in sample only.
BAYESIAN METHODS
Inference based on information from sample and prior subjective knowledge of the parameter
x x
PARAMETER ESTIMATION
TESTS OF HYPOTHESIS
A hypothesis is proposed regarding the value of a parameter or the relationship between parameters of two or more populations, and checked if it is true based on the evidence in the sample.
POINT ESTIMATION
The value of a population parameter is desired. A single value from one sample is proposed as an estimate for the parameter.
INTERVAL ESTIMATION
An interval is proposed and it is affirmed with a certain degree of confidence that the parameter lies in that interval.

FIGURE 3.1 · *Statistical methods for inference*

tor for μ, the population mean; and S or R can be used as estimator for σ of a population.

When a parameter is to be estimated there may be a few estimators available. Certain criteria are used to select a "good" estimator from the available estimators. An estimator is said to be *unbiased* if its average value (over several samples) is expected to equal the parameter being estimated. The notation E(.) is used to denote the expected value or the long-run average value of a random variable.

Using this notation, it can be shown $E(\overline{X}) = \mu$, $E(M) = \mu$, and $E(S^2) = \sigma^2$ where S^2 is the sample variance $= \dfrac{\Sigma(X_i - \overline{X})^2}{n - 1}$ Thus $\overline{X}$ and M are unbiased estimators for μ and S^2 is an unbiased estimator for σ^2.

Of all the unbiased estimators, the one with the least variance is called the most *efficient* estimator.

It can be shown that $\overline{X}$ and S^2 are such efficient or *minimum variance unbiased* estimators for μ and σ^2 respectively of normal populations.

So, if we are searching for an estimator to estimate μ and σ^2 we would select $\overline{X}$ and S^2, respectively, as they are such good estimators. This explains why these statistics are preferred to make inference about μ and σ^2 in the methods discussed herein.

A single value of an estimator is called a *point estimate*. The point estimate, however, is just one observation of the estimator which is a random variable. When we want to estimate a parameter, the point estimate is of no value unless the sample size is quite large. For the small samples that we usually have, the point estimate cannot be relied on to give the value of the parameter. Therefore, we resort to what are called *interval estimates* or *confidence intervals* (CIs).

Confidence Intervals

Suppose we want to estimate the mean μ of a population and we choose sample average $\overline{X}$ as the estimator. A sample from the population will be taken and the value of $\overline{X}$ will be computed from the sample observations, which will give a point estimate for μ. Using this value of $\overline{X}$, an interval $(\bar{x} - k, \bar{x} + k)$ will be created such that $P(\bar{x} - k \le \mu \le \bar{x} + k) = 1 - \alpha$. $(1 - \alpha)$ is called the confidence coefficient; the interval $(\bar{x} - k, \bar{x} + k)$ is called a $(1 - \alpha)100\%$ confidence interval for μ. The value of k can be determined from the knowledge of the distribution of the estimator $\overline{X}$.

The method of estimating population parameters through confidence interval is illustrated using a few examples. In these examples, the populations of interest are assumed to have the normal distribution.

CI for $\mu - \sigma$ Known

The sample average $\overline{X}$ is used as the estimator. On the assumption that the population has $N(\mu, \sigma^2)$, the $\overline{X}$ has $N(\mu, \sigma^2/n)$ where n is the sample size. Then,

$$\frac{(\overline{X} - \mu)}{\sigma/\sqrt{n}} \sim N(0, 1)$$

Using this result the CI for μ is given as follows:

A $(1 - \alpha)100\%$ CI for μ of a population that is normally distributed with a known standard deviation σ is given by:

$$[\bar{x} - z_{\alpha/2}\, \sigma/\sqrt{n}, \quad \bar{x} + z_{\alpha/2}\, \sigma/\sqrt{n}]$$

where $z_{\alpha/2}$ is the number such that $P(Z \geq z_{\alpha/2}) = \alpha/2$. Or $z_{\alpha/2}$ is the number that cuts off $\alpha/2$ probability at the upper tail of the standard normal distribution. See Figure 3.2.

Remember that this model can be used when the population can be assumed to be normally distributed and its standard deviation is known.

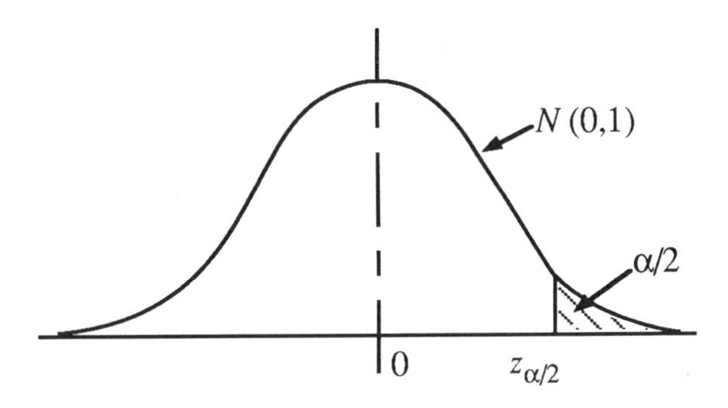

FIGURE 3.2 · *Definition of* $z_{\alpha/2}$

Example A random sample of four bottles of fabric softener from a filling line showed the following turbidity measurements: 12.6, 13.4, 12.8, 13.2 (in ppm). Find a 99% CI for the average turbidity for the population of bottles filled in this line. It is known that turbidity measurements are normally distributed with standard deviation = 0.3.

Solution

$$(1 - \alpha) = 0.99 \rightarrow \alpha = 0.01 \rightarrow \alpha/2 = 0.005 \rightarrow z_{\alpha/2} = 2.575$$

$$\bar{x} = 52/4 = 13.0$$

99% CI for μ:

$$[13.0 - 2.575(0.3/\sqrt{4}),\ 13.0 + 2.575(0.3/\sqrt{4})]$$

$$= [12.61,\ 13.39]$$

The CI can be interpreted as follows: If intervals are set up as above from sample of size 4, 99 out of 100 times such intervals will contain the true population mean.

CI for μ — σ Not Known

Suppose the population standard deviation is not known, a large sample ($n \geq 30$) can be taken and the sample standard deviation S from this large sample can be used as the known σ. If only a small sample is available, then a CI is created using the fact that the random variable $\frac{(\bar{X}-\mu)}{S/\sqrt{n}}$ has the t-distribution with $n - 1$ degrees of freedom.

The t-distribution is a symmetric distribution and its parameter is called the degrees of freedom (df). Its mean is zero. Its shape depends on the value of the parameter–df and approaches that of the standard normal distribution when the df becomes large, larger than 30. Tables are available that give the percentiles of the t-distribution with various degrees of freedom. A copy of the t table is provided in Appendix Table A.2.

A $(1 - \alpha)100\%$ CI for μ of a population that is normally distributed is given by:

$$[\bar{x} - t_{\alpha/2} (s/\sqrt{n}), \quad \bar{x} + t_{\alpha/2} (s/\sqrt{n})]$$

where $\bar{x}$ is the sample average and s is the sample standard deviation from a sample of size n, and $t_{\alpha/2}$ is such $P(t_{n-1} > t_{\alpha/2}) = \alpha/2$, i.e., $t_{\alpha/2}$ is the number that cuts off $\alpha/2$ probability at the upper tail of the t-distribution with df $= n - 1$. See Figure 3.3.

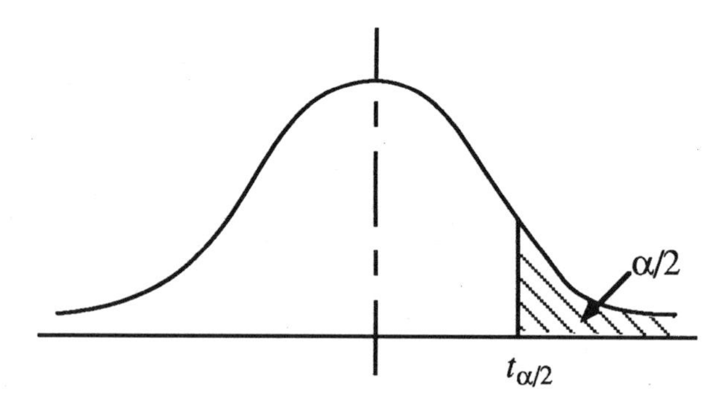

FIGURE 3.3 · *Example of a* t-*distribution*

Example

Four measurements of turbidity in bottles of fabric softener from a filling line are: 12.6, 13.4, 12.8, and 13.2. Set up 99% CI for average turbidity in the bottles of fabric softener. Assume normal distribution.

$$\alpha/2 = 0.005 \qquad n - 1 = 3 \qquad \bar{x} = 13.0$$
$$s = 0.365 \quad t_{0.005,3} = 5.841$$

$$\text{99 percent CI: } [13.0 \pm 5.841(0.365/\sqrt{4})] = [11.93, 14.07]$$

Note the CI is wider compared to the σ-known case in the previous example. It is less precise for lack of exact information on the population standard deviation.

CI for σ^2 of a Population

For setting CI for σ^2, the sample variance $S^2 = \dfrac{\Sigma(X_i - \bar{X})^2}{n - 1}$ is used as the estimator. The fact that the random variable $(n - 1)S^2/\sigma^2$ has the χ^2 (chi-squared) distribution with $(n - 1)$ degrees of freedom is used. That is,

$$\frac{(n-1)S^2}{\sigma^2} \sim \chi^2_{n-1}$$

The chi-squared distribution is a positive-valued distribution with a single parameter called the degrees of freedom (df). The shape of a χ^2 distribution depends on the value of the parameter df. Percentiles of the χ^2 distribution for various values of df are available in tables such as Appendix Table A.3.

A $(1 - \alpha)100\%$ CI for the variance σ^2 of a normal population is given by:

$$\left[\frac{(n-1)s^2}{\chi^2_{\alpha/2}}, \frac{(n-1)s^2}{\chi^2_{1-\alpha/2}}\right]$$

where s^2 is the sample variance, n the sample size and $\chi^2_{\alpha/2}$ is such that $P(\chi^2_{n-1} > \chi^2_{\alpha/2}) = \alpha/2$. See Figure 3.4.

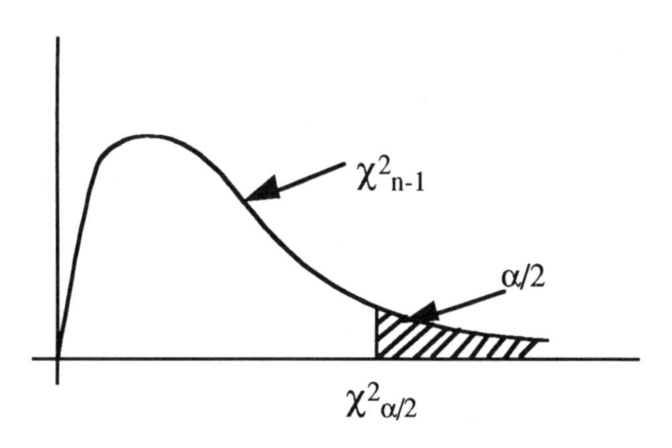

FIGURE 3.4 · *Example of chi-square distribution*

Example Set up a 99% CI for the standard deviation σ of turbidity in bottles of cloth softener, if a sample of four bottles gave measurements: 12.6, 13.4, 12.8, and 13.2. Assume normality.

Solution

$\alpha = 0.005$ $\qquad s^2 = 0.133$

$\chi^2_{0.005,3} = 12.84 \qquad \chi^2_{0.995,3} = 0.07$

$$99 \text{ percent CI for } \sigma^2 = \left[\frac{(3)(.133)}{12.84}, \frac{(3)(.133)}{0.07}\right]$$

$$= [0.031, 5.7]$$

$$99 \text{ percent CI for } \sigma = [\sqrt{.031}, \sqrt{5.7}] = [0.176, 2.39]$$

Note that the confidence interval for σ was obtained after first finding the confidence interval for σ^2.

Suppose we want a 95% CI for σ: $\chi^2_{.025,3} = 9.35$

$\chi^2_{.975,3} = 0.22$

95% CI for $\sigma^2 = [0.043, 1.82]$

95% CI for $\sigma = [0.207, 1.349]$

Note, the larger the confidence coefficient, the wider the CI becomes and vice versa.

Hypothesis Testing

Hypothesis testing is an inferential procedure where a hypothesis is first proposed about the value of a population parameter or the relationship between parameters of two (or more) populations. The hypothesis is then tested to check if it is true or not, based on the information obtained from sample(s) taken from the population(s).

The hypothesis proposed is called the *null hypothesis* denoted by H_0. An *alternate hypothesis* denoted by H_1 is also proposed along with H_0, which will be accepted if H_0 is found not acceptable.

After the hypotheses have been set up, a sample is taken from the population(s) in question and the value of an appropriate *test statistic* is computed from the sample.

In any statistical procedure there are two possible types of errors. When a statistical method is used there are four possible outcomes as shown in Figure 3.5.

Type-I Error is said to have occurred if H_0 is declared false when in

		Test declares H_0	
		True	False
H_0 is in reality	True	OK	Error Type I
	False	Error Type II	OK

FIGURE 3.5 · *Possible outcomes when a test is performed*

fact it is true. Type-II Error is said to have occurred if H_0 is declared true when in fact it is false.

When a statistical procedure is used, the probability of these errors can be calculated from the knowledge of the distribution of the test statistic. More importantly, the test procedure can be designed in such a way that the probability of an error occurring can be contained within a specified value.

Probability of Type I error is denoted by α and is also called the *level of significance* or *size of the test*. Probability of Type II error is denoted by β.

Designing the test involves setting up a *critical region* in the distribution of the test statistic such that if the observed value of the test statistic (value computed from sample) falls in that region the null hypothesis will be rejected in favor of the alternate hypothesis; otherwise the null hypothesis will not be rejected.

The steps involved in hypothesis testing can be summarized as follows:

1. Set up H_0 and select an appropriate H_1.
2. Choose an appropriate test statistic.
3. Choose a level of significance.
4. Design the test by specifying the critical region (CR).
5. Select a sample from the population and compute the value of the test statistic.
6. Reject H_0 if the observed value of the test statistic is in CR; otherwise do not reject.

The following examples illustrate the method of using hypothesis testing to draw inference about population parameters.

Test Concerning the Mean $\mu - \sigma$ Known

H_0: $\mu = \mu_0$ (Hypothesize that the mean equals a number μ_0)

H_1: $\mu < \mu_0$ (The alternate hypothesis is: If the mean is not equal to μ_0 it must be less than μ_0)

$$\text{Test Statistic: } \frac{\overline{X} - \mu_0}{\sigma/\sqrt{n}} \sim Z$$

The test statistic relates the sample average to the population mean and it has a distribution that is known to be $N(0, 1)$.

Critical Region: All observed values of Z that are less than $-z_\alpha$. That is, reject H_0 if the observed value of Z, $z_{obs} < -z_\alpha$, otherwise do not reject H_0.

The critical region in this case determines how small the value of the test statistic should be before the null hypothesis should be rejected in

Case 1 $H_0: \mu = \mu_0$
$H_1: \mu < \mu_0$

Case 2 $H_0: \mu = \mu_0$
$H_1: \mu > \mu_0$

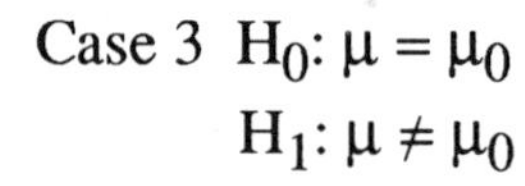

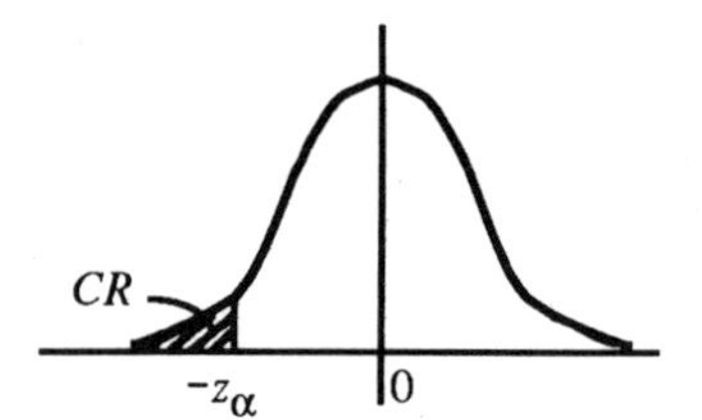

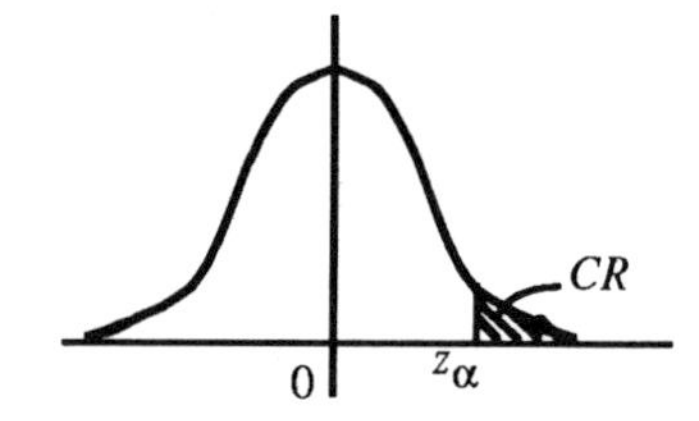

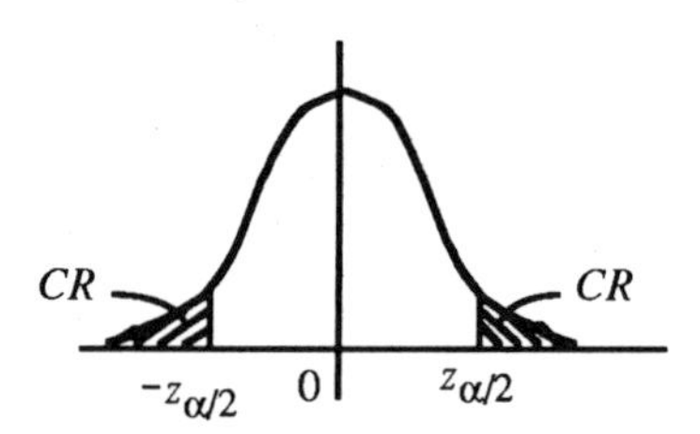

FIGURE 3.6 · *Alternate hypotheses and corresponding critical regions*

favor of the alternate hypothesis. When we use this critical region it should be noticed, there is the probability α that the null hypothesis will be rejected when in fact it is true.

Example

A supplier of a special rope claims their new product has average strength not less than 10 kg with standard deviation = 0.5 kg. A sample of 50 rope lengths gave an average of 9.8 kg. Test the hypothesis $\mu = 10$ vs. $\mu < 10$, with $\alpha = 0.01$.

Solution

H_0: $\mu = 10$
H_1: $\mu < 10$

$$\text{Test Statistic: } \frac{\overline{X} - \mu_0}{\sigma/\sqrt{n}} \sim Z$$

Critical Region: $Z_{obs} < -2.327$, i.e., values of Z_{obs} less than -2.327 will lead to rejection of H_0

$$z_{obs} = \frac{9.8 - 10}{0.5/\sqrt{50}} = -2.828$$

The computed value of z_{obs} is in the CR. So, this test leads to rejection of H_0. The average strength of the ropes in question *is* less than 10 kg.

This is an example of a *one-tailed test* because the critical region is on one tail of the distribution as shown in Figure 3.6 Case 1.

The type of alternate hypothesis determines where in the distribution of the test statistic the CR is defined. There are three possible alternate hypotheses encountered in testing situations. The test statistic will be the same for all three situations; however the location of CR will differ as shown in Figure 3.6.

In Case 1 we know from the facts of the situation that μ will be less than μ_0 if it is not equal to μ_0, and the CR determines how *small* the observed value of the test statistic should be if the hypothesis $\mu = \mu_0$

Case 1 $H_0: \mu = \mu_0$
$H_1: \mu < \mu_0$

Case 2 $H_0: \mu = \mu_0$
$H_1: \mu > \mu_0$

Case 3 $H_0: \mu = \mu_0$
$H_1: \mu \neq \mu_0$

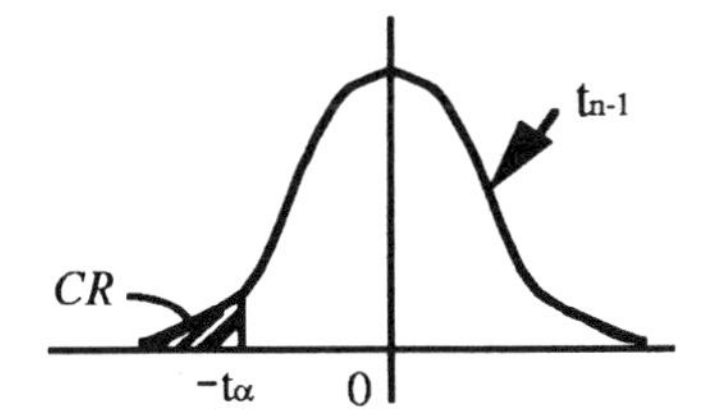

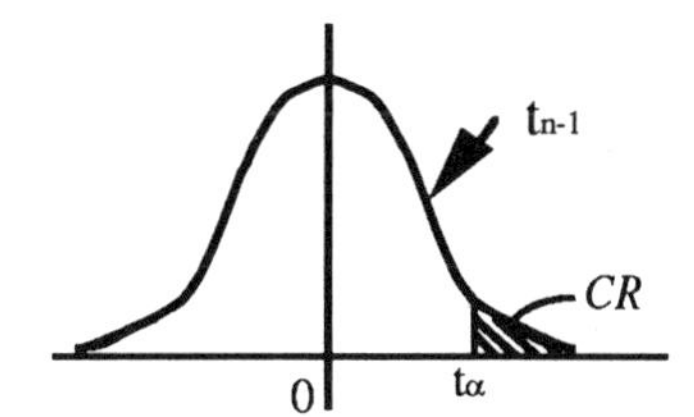

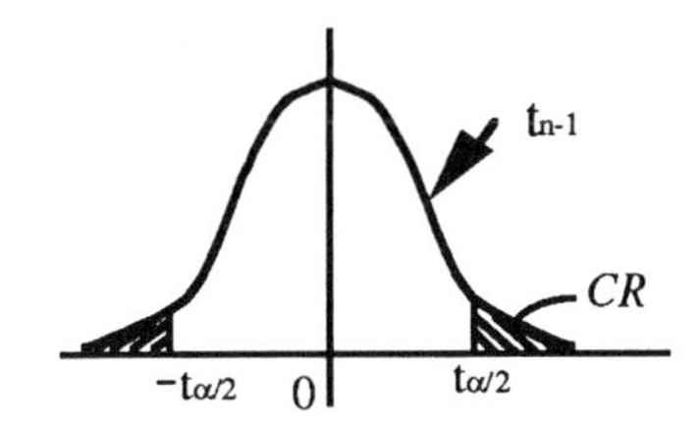

FIGURE 3.7 · *Critical regions for tests for μ when σ is not known*

should be rejected in favor of the alternate. There is still a risk of α that the null hypothesis will be rejected when in fact it is true.

In Case 2 we know from the facts of the situation that μ will be greater than μ_0 if it is not equal to μ_0, and the CR determines how *large* the value of the test statistic should be before the null hypothesis is rejected in favor of the alternate.

In Case 3 we have no specific information as to whether μ will be smaller or larger than μ_0 if it is not equal to μ_0, and the CR determines how far away from 0 the test statistic should be before the null hypothesis is rejected. Here the risk of α is divided equally on either side of the distribution since the error could occur on either side.

Case 1 is referred to as the *left-tailed test*; Case 2 the *right-tailed test*; and Case 3 the *two-tailed test*, based on location of the CR in the distribution of the test statistic.

Test Concerning the Mean $\mu - \sigma$ Not Known

As stated earlier, there are again three possible alternate hypotheses as shown in Figure 3.7.

$$\text{Test statistic: } \frac{\overline{X} - \mu_0}{S/\sqrt{n}} \sim t_{n-1}$$

When σ is not known we use the sample standard deviation S in the test statistic. This test statistic is known to have the t-distribution. The critical regions corresponding to the three alternate hypotheses are shown in Figure 3.7.

Example The amount of protein per box of cereal should be more than two grams according to a manufacturer's advertisement. Lab analysis of five boxes gave the following results: 1.8, 2.1, 2.15, 2.02, and 1.96 grams. Does the sample show evidence that the average protein content in the boxes produced by the manufacturer is more than two grams? Use $\alpha = 0.05$; assume normality.

Solution H$_0$: $\mu = 2$

H$_1$: $\mu > 2$

Notice that the conclusion that we want to draw as a result of the test is in the alternate hypothesis. This is because a conclusion drawn as a result of rejecting a null hypothesis is stronger and more dependable.

$\bar{x} = 2.006 \qquad s = 0.1363$

Test statistic: $\dfrac{\bar{X} - \mu_0}{S/\sqrt{n}} \sim t_4$

Critical region: Observed values of $t > t_{0.05,4}$
The observed value of the test statistic:

$$t_{\text{obs}} = \frac{2.006 - 2}{0.1363/\sqrt{5}} = 0.0984$$

$t_{0.05,4}$ from Table A.2 $= 2.132$
t_{obs} is not in CR $\Rightarrow$ Do not reject H$_0$ $\Rightarrow$ Average protein content is *not* more than 2 grams. The manufacturer's claim is not valid.

Test for Difference of Two Means — σs Known

H$_0$: $\mu_1 - \mu_2 = 0$; i.e., no difference between the two population means.

There are again three possible alternate hypotheses.

Case 1	Case 2	Case 3
H$_0$: $\mu_1 - \mu_2 = 0$	H$_0$: $\mu_1 - \mu_2 = 0$	H$_0$: $\mu_1 - \mu_0 = 0$
H$_1$: $\mu_1 - \mu_2 < 0$	H$_1$: $\mu_1 - \mu_2 > 0$	H$_2$: $\mu_1 - \mu_2 \neq 0$

Test statistic: $\dfrac{(\bar{X}_1 - \bar{X}_2)}{\sqrt{\sigma_1^2/n_1 + \sigma_2^2/n_2}} \sim Z$

The critical regions corresponding to the three alternate hypotheses are shown in Figure 3.8.

Case 1 H$_0$: $\mu_1 - \mu_2 = 0$
H$_1$: $\mu_1 - \mu_2 < 0$

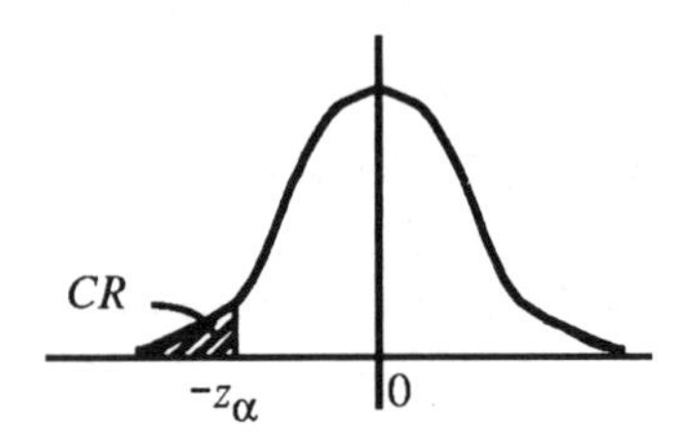

Case 2 H$_0$: $\mu_1 - \mu_2 = 0$
H$_1$: $\mu_1 - \mu_2 > 0$

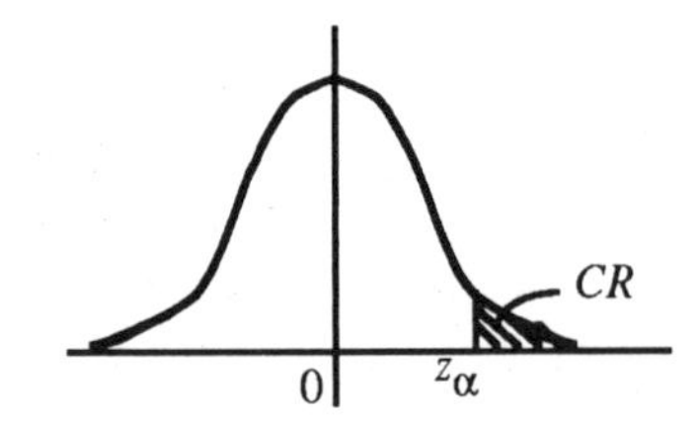

Case 3 H$_0$: $\mu_1 - \mu_2 = 0$
H$_1$: $\mu_1 - \mu_2 \neq 0$

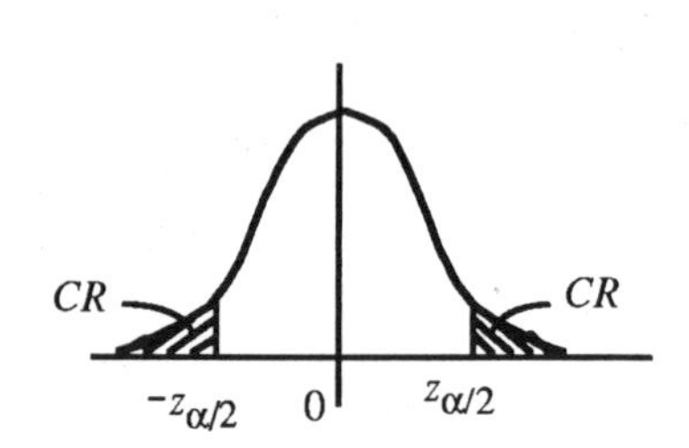

FIGURE 3.8 · *Critical regions for tests for difference of two means when σ's are known*

Example

A random sample of 100 male workers in a factory had an average salary of \$35,000 with standard deviation of \$1200. Another random sample of 80 female workers had an average of \$34,600 with standard deviation of \$1800. Test the hypothesis that male and female workers in this factory earn equal pay against the alternate that they don't, at $\alpha = 0.01$.

Solution

MALE WORKERS	FEMALE WORKERS
$\bar{x}_1 = 35{,}000$	$\bar{x}_2 = 34{,}600$
$\sigma_1 = 1200$	$\sigma_2 = 1800$
$n_1 = 100$	$n_2 = 80$

Since n_1 and n_2 are both large, the sample standard deviations are used as the known population standard deviations.

H_0: $\mu_1 - \mu_2 = 0$

H_1: $\mu_1 - \mu_2 \neq 0$

The two-sided alternate hypothesis is chosen because there is no reason to believe the difference in average will be positive or negative; it could be either way.

$$\text{Test statistic: } \frac{(\bar{X}_1 - \bar{X}_2)}{\sqrt{\sigma_1^2/n_1 + \sigma_2^2/n_2}} \sim Z$$

$$z_{obs} = \frac{400}{\left[\dfrac{1200^2}{100} + \dfrac{1800^2}{80}\right]^{1/2}} = \frac{400}{234.3} = 1.71$$

$$z_{\alpha/2} = z_{0.005} = 2.575$$

Critical region: Observed values of $Z < -2.575$ or > 2.575.

z_{obs} is not in the critical region, therefore, do not reject H_0. There is no evidence to show that there is a difference between the salaries of male and female workers.

We have examined only a few basic models in hypothesis testing just to get an idea of where and how we can use this statistical procedure. There are models that can be used to test hypotheses for difference of two means when population standard deviations are not known and sample sizes are small. We then use a test statistic that has *t*-distribution. The model to test the hypothesis about population variance uses a test statistic that has χ^2-distribution. The model to test the hypothesis about ratio of two variances uses a test statistic that has *F*-distribution. All of these models assume that populations concerned have normal distribution.

There are testing procedures that do not require normality assumption. These are known as distribution-free tests or nonparametric tests.

There also are tests for testing the nature of the distribution of a population known as goodness of fit tests. Details of all these tests can be found in any statistics book. The goodness of fit test as it relates to exponential distribution is discussed in Chapter 5.

EXERCISE

3.1 A random sample of 12 specimens of core sand in a foundry has a mean tensile strength of $\bar{x} = 180$ psi. Construct a 95% CI on mean tensile strength of core sand under study. Assume the tensile strength is normally distributed and the standard deviation is known to be 1.5 psi.

3.2 A random sample of 20 deodorant cans were found to have a mean content of $\bar{x} = 1.15$ ounces of concentrate. The standard deviation of the contents $s = .025$ ounce. Find a 90% CI on the mean quantity of concentrate in each can of deodorant.

3.3 A random sample of 15 steel rods gave the following measurements of diameter in millimeters. Assuming that the rod diameter is normally distributed, construct a 95% CI for the mean and variance of the rod diameter. What is the 95% CI for the standard deviation?

6.24, 6.23, 6.20, 6.21, 6.20, 6.28, 6.23, 6.26, 6.24, 6.25, 6.19, 6.25, 6.26, 6.23, 6.24.

3.4 An automatic core making process is being studied in a foundry. The past six shifts of plant operation have resulted in the following yields (in percentages): 92, 86, 91, 90, 91, and 86. Is there reason to believe that the yield is less than 90 percent? Assume $\alpha = 0.05$ and the standard deviation of yield is known to be 5 percent.

3.5 The life of a battery is of interest to a manufacturer. A random sample of eight batteries gave the following lives in months: 108, 138, 124, 163, 124, 159, 106, and 134. Is there any evidence that the mean life is greater than 125 months? Assume that battery life is normal. Use $\alpha = 0.05$.

3.6 Two machines are used for filling bottles with a type of liquid soap. The filled volumes can be assumed to be normal, with standard deviations $\sigma_1 = .12$ and $\sigma_2 = .16$. A random sample is taken from the output of each machine and the following weight checks were obtained. Check if both machines are filling equal volumes in the bottles. Use $\alpha = 0.05$.

Machine 1: 18.03, 18.05, 18.04, 18.05, 18.02
18.01, 17.95, 17.99, 18.02, 17.92

Machine 2: 17.02, 18.03, 17.97, 18.04, 17.96
18.02, 18.01, 18.01, 17.99, 17.03

PART II

Reliability Prediction, Estimation, and Apportionment

CHAPTER 4

Reliability Concepts

Definitions

Reliability

Reliability is defined as the probability that a given product will successfully perform a required function without failure, under specified environmental conditions, for a specified period of time.

A reliability measure is needed to answer questions such as: "How long will the product last without breakdown?" "What proportion of the population of the product will fail before the warranty period?" "How long a warranty can be given for a new product?" In any given product population, some units may last 100 hours, some 200 hours, and others 2000 hours; in other words the life length is a random variable. When questions must be answered regarding the behavior of such random variables, it is necessary to use probability, distributions, averages and measures of variability. Reliability, being a function of this random variable, is therefore expressed as a probability; and statistics is used in its computation and prediction.

It must be realized, while defining the reliability of a product, the function the product is expected to perform must be clearly specified along with a definition of what constitutes a failure. An electric bulb will be considered to have failed when it burns out, an electric motor may be considered to have failed when it draws too much current, and a luxury car may be considered to have failed when the ride becomes a little rough. It is necessary to define exactly what function the product is expected to perform and what would be considered a failure, before its life can be measured.

Further, the conditions under which a product is expected to perform the required function must also be clearly specified. Reliability under one set of operating conditions may be different from that under another.

Finally, reliability must be expressed as a function of time. At any specified time a certain proportion of the product population will continue to successfully perform the required function without failure. Reliability can be interpreted to represent this proportion of the population that survives beyond the specified time.

Life Distribution

The life distribution of a product is the basic information from which all measures of reliability are evaluated. It is the distribution of length of life of all items in the population of a product. The length of life can be measured in hours, weeks, number of cycles, number of miles, etc. The distribution can be "estimated" from a set of sample life data taken from the population. Such data can be generated by testing a sample in the laboratory or observing it in actual field use. Oftentimes, it is possible to assume that the life distribution has a certain shape. A common assumption is that the life is exponentially distributed, in the same way most measurements arising from inspection are assumed to be normally distributed. Such assumptions, however, must be based on knowledge of the history of the product or of similar products.

The life distribution of a product is known if its shape such as exponential, normal, etc., is known along with estimated value(s) for the parameter(s) of the distribution. Parameters of a distribution, it should be recalled, are those quantities that, if known, completely describe the population to which they pertain.

Failure Rate or Hazard Rate

Failure rate represents the proneness to failure of a product as a function of its age, or time in operation.

Failure rate at any given time is the proportion that will fail in the next unit of time, out of those units that have survived up to that time. This requires further explanation.

Take, for example, 1000 electric motors put on test at time 0. Four hundred of them are working at 2000 hrs, 50 of them fail in the next 100 hrs, and another 50 fail in the following 100 hrs. See Figure 4.1.

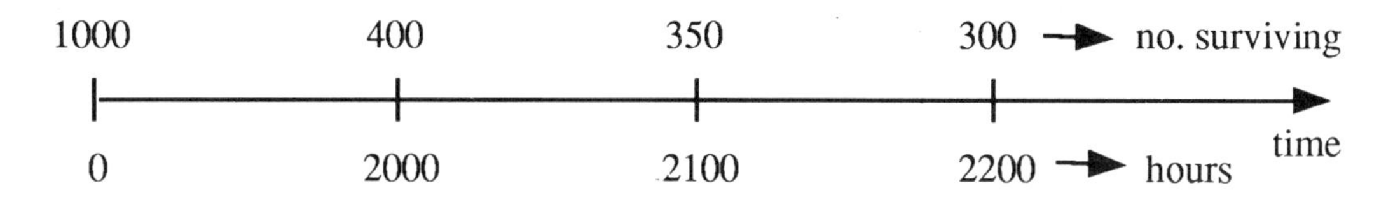

FIGURE 4.1 · *Failure data for electric motors*

The failure rate for the motors at 2000 hrs:

$$h(2000) = \frac{\text{(no. failing per hour following 2000 hrs)}}{\text{no. surviving at 2000 hrs}}$$

$$= (50/100)/400 = 0.00125 \text{ units/hr.}$$

Similarly, the failure rate at 2100 hr:

$$h(2100) = (50/100)/350 = 0.0014 \text{ units/hr}$$

(The $h(.)$ function is used to denote failure rate or hazard rate.)

The failure rate can increase, decrease, or remain constant over time depending on the product's nature. Failure rate can be evaluated from the knowledge of the life distribution. Average failure rate is used as a measure of reliability of a product. For products with exponentially distributed life, failure rate which remains constant over time, provides all the information needed about reliability. How the failure rate changes over time gives an insight into the failure mechanisms of a product and is used in studies for improving reliability.

The following example illustrates the method of evaluating reliability and failure rate of a product from sample data. It also helps in further understanding the definition of these terms.

Example

Table 4.1 shows the result of life test of 1000 refrigerator compressors giving the number of failures in different time intervals. Draw the life distribution, reliability, and failure rate curves as functions of time.

The calculations for reliability and failure rate are shown in Table 4.1. For example, the failure rate of the compressors that are 10 months old are calculated as:

$$h(10) = \frac{\text{(no. failing in one mo. after 10th mo.)}}{\text{no. surviving @ the end of 10th mo.}}$$

$$= (61/10)/653 = 0.0093 \text{ units/mo.}$$

This means that about 0.93% of the 10-month-old compressors will fail within a month. This failure rate increases to about 10% for the 80-month-old compressors. Also, these compressors had a larger tendency to fail when they were newer, less than 10 months old.

(1) INTVL MONTH	(2) NO. OF FAILURES	(3) % FAIL IN INTVL	(4) NO. OF SURVRS @ END OF INTVL	(5) RELIAB @ END OF INTVL	(6) FAIL RATE @ END OF INTVL
0–0	0	0.0	1000	1.000	0.0347
$0<t\leq 10$	347	34.7	653	0.653	0.0093
$10<t\leq 20$	61	6.1	592	0.592	0.0117
$20<t\leq 30$	69	6.9	523	0.523	0.0166
$30<t\leq 40$	87	8.7	436	0.436	0.0232
$40<t\leq 50$	101	10.1	335	0.335	0.0307
$50<t\leq 60$	103	10.3	232	0.232	0.0435
$60<t\leq 70$	101	10.1	131	0.131	0.0740
$70<t\leq 80$	97	9.7	34	0.034	0.1000
$80<t\leq 90$	34	3.4	0	0.00	-.---
Total	1000	100.0			

TABLE 4.1 · *Life distribution, reliability, and failure rate calculations for air compressors*

Note that column (3) gives the frequency distribution of failures and column (6) the failure rate. While the column (3) figures are percentages failing out of the original numbers, the column (6) represents proportion failing out of the units that have survived up to that time. Column (5) gives the reliability, being the proportion surviving at that time. Figure 4.2 shows the graph of failure distribution, reliability, and failure rate for the compressors.

The concept of failure rate is an important one in reliability studies. It helps in understanding the susceptibility of parts and equipment to fail when they are new, when they are broken in, and when they become old. Such understanding helps in planning reliability improvement strategies, preventive maintenance, parts replacement, and parts inventory planning.

The Bathtub Curve

The life behavior of many different types of equipment has been studied using failure rate curves. Figure 4.3 shows a few different curves showing how the failure rate changes over time for different types of equipment. The one failure rate curve that seems applicable to a wide variety of complex equipment is shown in Figure 4.3(d). Because of its shape, it is known as the *bathtub curve*. Such a curve divides the life of equipment into three distinct regions.

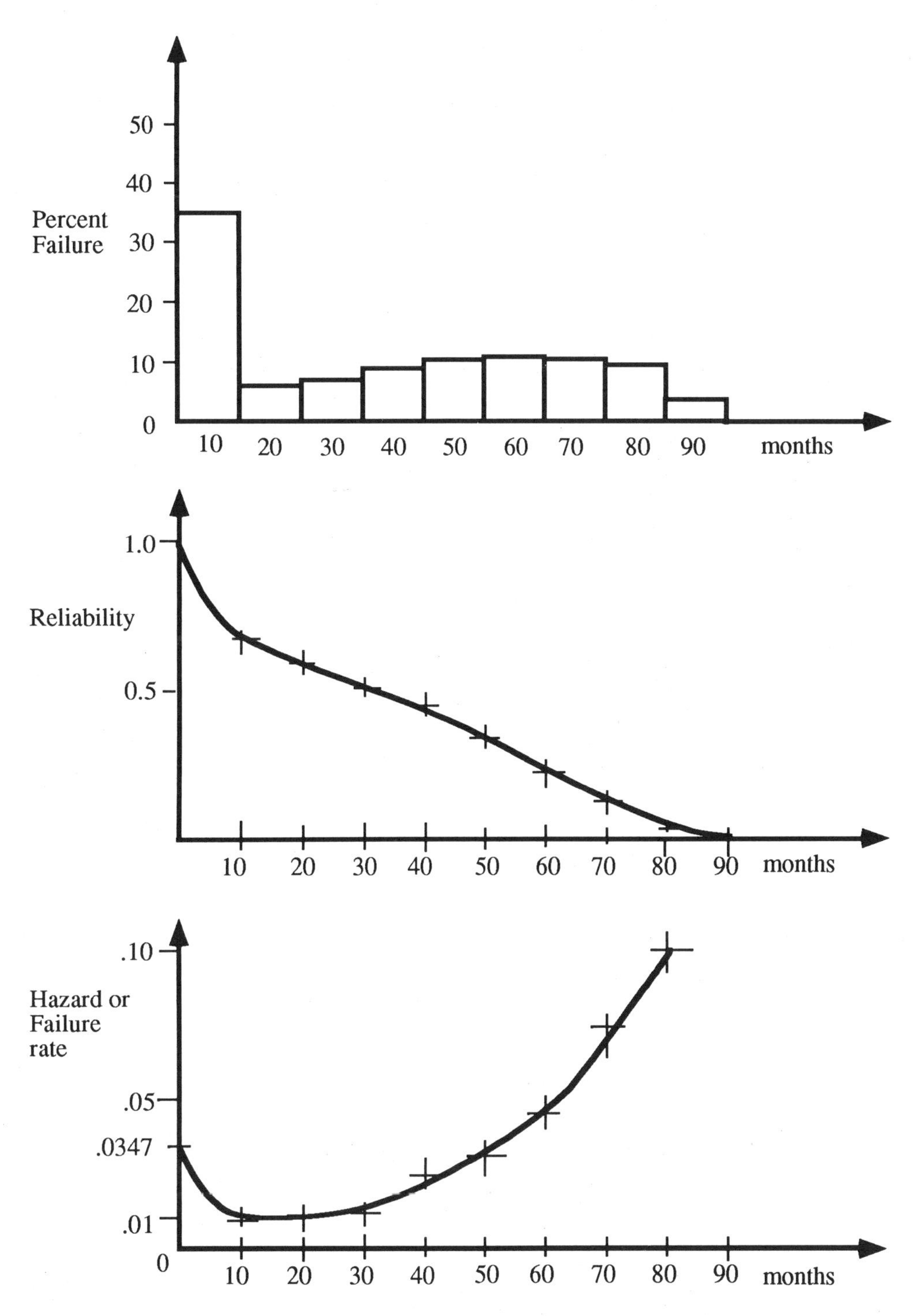

FIGURE 4.2 · *Life distribution, reliability, and failure rate functions of air compressors (data in Table 4.1)*

A. Region of decreasing failure rate until the rate becomes constant. The decrease in failure rate is due to defective units in the population—defective due to poor material or bad workmanship—failing early and being repaired or removed from the population. This

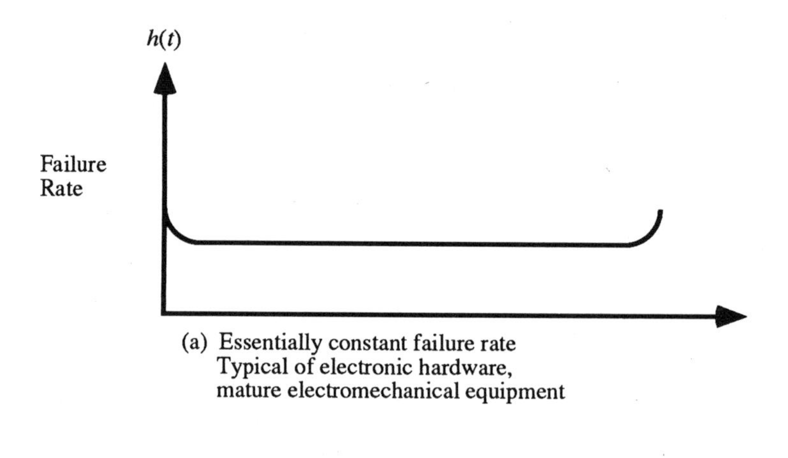

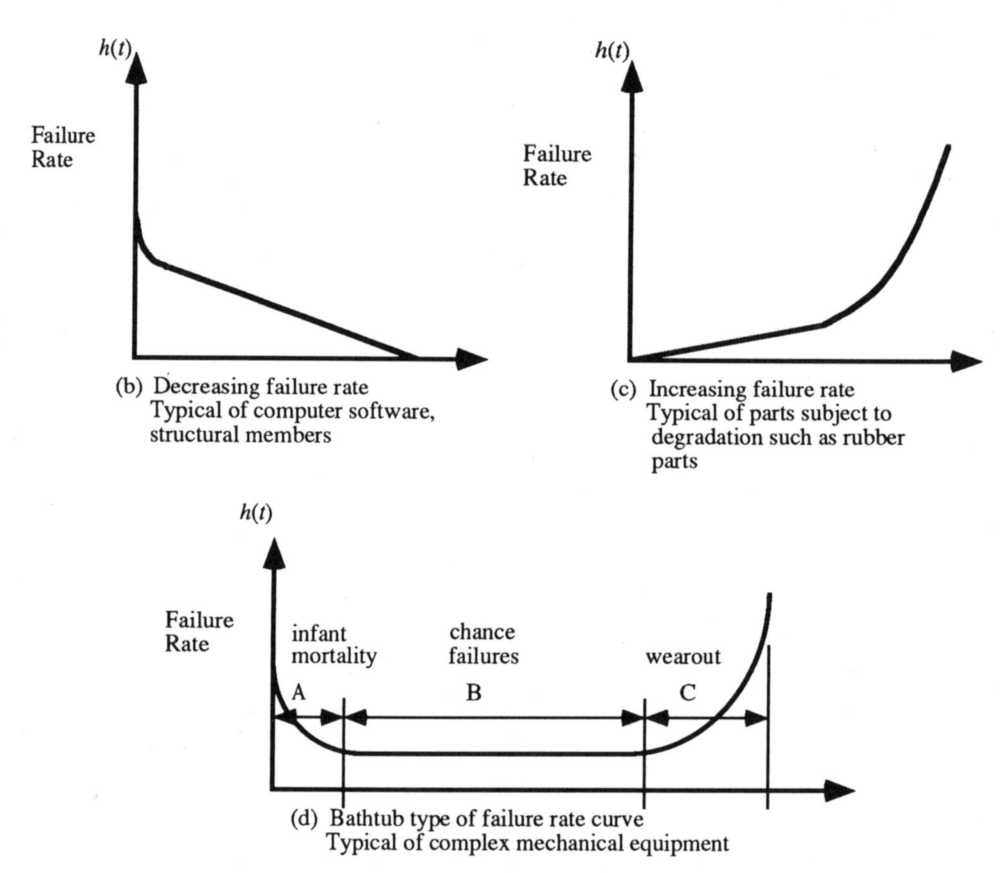

FIGURE 4.3 · *Different types of failure rate curves*

region is known as the *infant mortality* region. A high initial failure rate or long period for infant mortality would indicate inadequate quality control effort for parts and assemblies. The length of this period determines how long a burn-in is required.

B. Region of constant failure rate until the rate starts increasing. Here, failures occur not because of inherent defects in the units but because of accidental or chance occurrence of loads in excess of the design strength. As will be seen later, constant failure rate implies that the life distribution of units in this region follows exponential law. This region is known as the *period of chance failures.*

C. Region of increasing failure rate because of parts wearing out. Failures occur due to fatigue, aging, and embrittlement. This region is known as *wearout period.* Knowledge of when wearout begins helps in planning replacements and overhauls.

Mean Time to Failure (MTTF)

The MTTF is the average or mean of the life distribution. In other words, it is the average of life length of all the units in the population. The term MTTF is normally used for products that have only one life, i.e., not repairable. For products that are repairable, the term mean time between failures (MTBF) is used to denote the average time between repairs. The MTTF (or the MTBF) has a special significance when the life distribution is exponential because, then it is equal to the reciprocal of the parameter of the distribution, the (constant) failure rate. So knowledge of the MTTF equates to knowledge of the entire distribution. Evaluation and prediction of all measures relating to reliability can then be made with MTTF only.

It should be pointed out, however, that the MTTF and MTBF do not have the same importance when the life distribution is other than exponential. For example, the Weibull distribution can be used to model a variety of life distributions with decreasing, increasing, and constant failure rates. When the Weibull is used to model distributions other than the constant failure rate case, knowledge of MTTF or MTBF alone is not adequate. Then the parameters of the Weibull must be estimated.

EXERCISE

4.1 Find the life distribution, reliability, and failure rate functions of the data shown below, which represent life in hours of turn signals. Plot the functions on a graph paper.

Time	No. failed
0–20	20
20–40	16
40–60	12
60–80	10
80–100	5
100–150	17
150–200	7
200–300	10
300–400	3
>400	0
	100

CHAPTER 5

Mathematical Prediction and Estimation of Reliability

From the definitions in the previous chapter it can be seen that reliability is quantified using probability and reliability of an equipment can be evaluated if its life distribution is known. In Table 4.1 the method of calculating reliability from observed data from samples was shown. Now the method of calculating reliability measures based on mathematical models for life distribution will be discussed.

The Exponential Distribution

In many situations the life distribution can be expected to follow the exponential law, especially if the product has matured beyond the infant mortality stage and wearout has not set in. If data are collected on life length of products in this period, the histogram will appear as in Figure

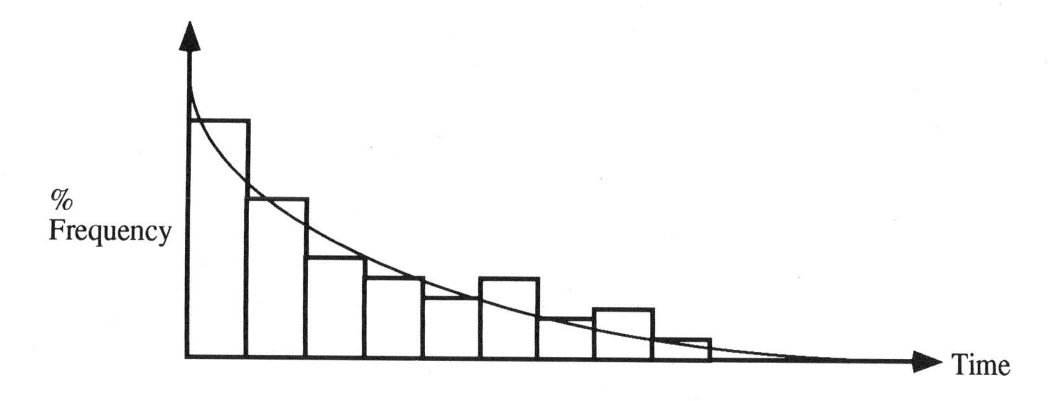

FIGURE 5.1 · *Histogram of exponential data*

5.1 characterized by a concentration of values near zero and a tapering off with a long tail.

If the histogram can be approximated by a smooth curve, this curve can be represented by a mathematical expression which is called the exponential probability density function.

The probability density function (pdf) can be interpreted as a mathematical expression which approximates frequency distributions of populations. When a life distribution can be represented by a mathematical expression, quantities such as reliability and failure rate can be derived from it mathematically.

The exponential pdf is expressed as a function of time t in the form:

$$f(t) = \lambda e^{-\lambda t}, t \geq 0.$$

In this expression λ is a constant that takes different values for different populations and is the parameter of the distribution. When the exponential is used as a model for the life of a product the quantity λ represents, as will be seen below, the product's failure rate.

Once it is known that the life of a product follows this distribution, reliability predictions about the life of the product can be made from this mathematical model.

Figure 5.2(a) shows the graph of an exponentially distributed life time. If the total area under the curve represents 100% of the population, the area beyond a time t represents the proportion that would survive beyond t, which is the reliability at t. This area can be evaluated using integration as:

$$R(t) = \int_t^{\infty} \lambda e^{-\lambda x} dx = e^{-\lambda t}$$

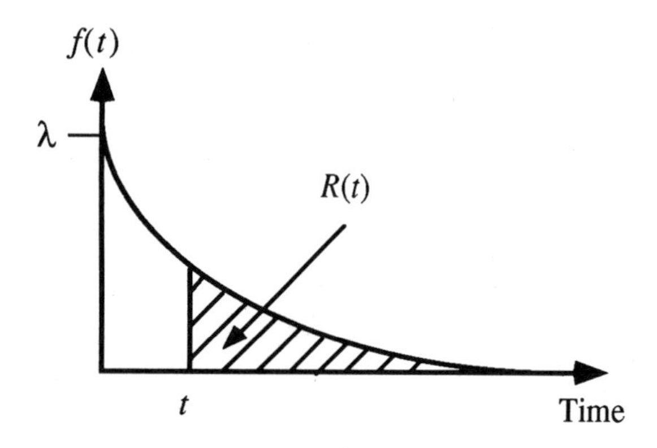

FIGURE 5.2 · *Graph of exponential pdf*

For any population that has life distribution represented by a pdf, $f(t)$, the failure rate $h(t)$ at time t is given by the expression:

$$h(t) = \frac{f(t)}{R(t)}$$

It is easy to see that this expression will indeed give the failure rate. The numerator represents the rate at which units of the population fail in the instant following time t and the denominator the proportion of units surviving at time t.

So, for exponential distribution, the failure rate

$$h(t) = \frac{f(t)}{R(t)} = \frac{\lambda e^{-\lambda t}}{e^{-\lambda t}} = \lambda, \text{ a constant.}$$

Therefore, if a product has exponential life distribution its failure rate does not change over time and the constant failure rate is equal to the parameter of the distribution, λ. The converse of this statement also is true, i.e., if an equipment has constant failure rate its life must be exponential.

It also can be shown (using the definition for the mean of a distribution given in Chapter 2) that if life is exponentially distributed the mean life of the population $= \frac{1}{\lambda}$. That is, it can be shown that:

$$\int_0^\infty t\lambda\, e^{-\lambda t} dt = \frac{1}{\lambda}$$

Therefore, for exponential life:

$$\text{MTTF (or MTBF)} = \frac{1}{\text{failure rate}}$$

In summary, if a product's life distribution can be approximated by the exponential distribution:

Life distribution (pdf): $f(t) = \lambda e^{-\lambda t}$

Reliability at t: $R(t) = e^{-\lambda t}$

Failure rate: $h(t) = \lambda$

Mean life: $\text{MTTF (or MTBF)} = \dfrac{1}{\lambda}$

Example

A certain type of engine seal is known to have life exponentially distributed with a (constant) failure rate $= 0.03 \times 10^{-4}$ failures per hour.

a) What is the probability a given seal will last beyond 10,000 hours?

b) What is the MTTF of the seal?

c) What is the reliability at MTTF?

d) If the reliability at design life has to be at least 90%, what is the recommended design life?

Solution

$\lambda = 0.03 \times 10^{-4}$ per hour

a) Need reliability at 10,000 hrs

$R(t) = e^{-\lambda t}$

$R(10000) = e^{-.000003 \times 10000} = e^{-0.03} = 0.97$

i.e., 97 percent of the seals will last beyond 10,000 hours.

b) $\text{MTTF} = \dfrac{1}{\lambda} = \dfrac{1}{(0.03 \times 10^{-4})} = 333{,}333$ hours

i.e., the average life of these seals is 333,333 hours.

c) $R(333333) = e^{-0.000003 \times 333333} = 0.368$

i.e., only 37 percent of the seals will last beyond the average life.

d) Let T be the recommended design life. For 90 percent reliability,

$R(T) = 0.9;$ i.e., $e^{-0.000003T} = 0.9$

Taking logarithm of both sides,

$-0.000003 \times T = -0.1054$

$T = 35{,}133$ hrs.

It should be noted from the above example that if the life distribution of a product can be assumed exponential, an estimate for the failure rate alone is enough to answer several questions on the product's life. Reliability of products with exponential life at MTTF is only 0.368; i.e., only 36.8% of such population will survive beyond the average life. If design life has to be chosen with reasonably large reliability, it will have to be much smaller than the MTTF.

How does one ascertain if the life of a part or equipment is exponentially distributed? How can the parameter, the constant failure rate, be estimated? Answers to these questions must come from data collected from samples. There are certain peculiarities associated with data coming from life tests. These will be discussed first.

Types of Data

While examining data on product life it should be recognized that there are different kinds of them. In regular quality control work if a sample of 10 items is inspected, 10 observations will be obtained. Such data are known as *complete data.* In life testing, when a sample of 10 is put on test, very rarely will 10 observations become available, because some of the items in the sample may not fail within a reasonable period of time and the test may have to be stopped before all sample units fail. Under these circumstances or when an analysis is desired at an intermediate stage before the test is completed, *incomplete data* or *censored data* will result.

Censored data can be further classified into three types: singly censored-Type I, singly censored-Type II, and multicensored. It is necessary to understand what type of data one has in order to correctly analyze them.

Singly Censored—Type I

Figure 5.3(a) shows the conditions that generate this type of data. The test is stopped at a predetermined time T. The data are called singly censored because all survivors or runouts are removed from the test at the same time. When the runouts have different runout times, as it happens under certain experimental or field use conditions, the data are said to be multicensored.

Singly censored—Type I data often is referred to simply as *time-censored* or *time-truncated* data.

Singly Censored—Type II

Figure 5.3(b) shows the circumstances under which this type of data arise. The test is stopped as soon as a predetermined number of failures have occurred. All runout units have the same runout times and they are equal to the failure time of the last failure.

The singly censored—Type II data are simply called *failure-censored* or *failure-truncated* data.

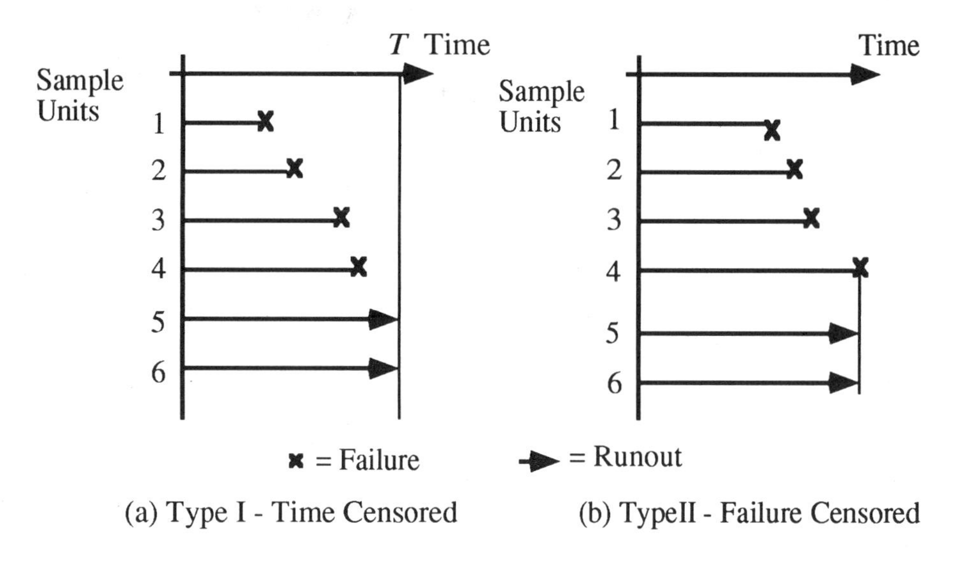

FIGURE 5.3 · *Singly censored data*

Multi-Censored

Multicensored data are characterized by the runout units having different runout times. Such data can arise from different situations. Figures 5.4(a) and (b) show two examples. In Figure 5.4(a), units 1, 2, and 3 were sold to customer 1 and when failure of unit 1 was reported, others were working. Units 4, 5, and 6 were sold to customer 2 and when the failure of unit 4 was reported, units 5 and 6 were still working. In Figure 5.4(b), out of six units put on test, three failed at the times shown, but for the other three, the test fixtures failed before the units failed. So the units had to be removed from the test at different times when the fixtures failed.

There are many other situations in which such multicensored data arise. Analysis methods for such data, as also for the singly censored data, will include the information that those units that did not fail were still running at the time they had to be taken out of the test. Such information usually adds to the precision or confidence in the results.

Interval Data

Yet another type of data encountered in reliability studies is called the interval data. Figure 5.5 depicts the situation where interval data arise. It is only known that certain sample units failed in certain time intervals, their exact time of failure being unknown. This occurs when the samples are inspected at specified times and their conditions noted. This type of data comes from both field and laboratory tests.

It is necessary to apply the correct type of analysis for a given type of data in order to get the most information from it.

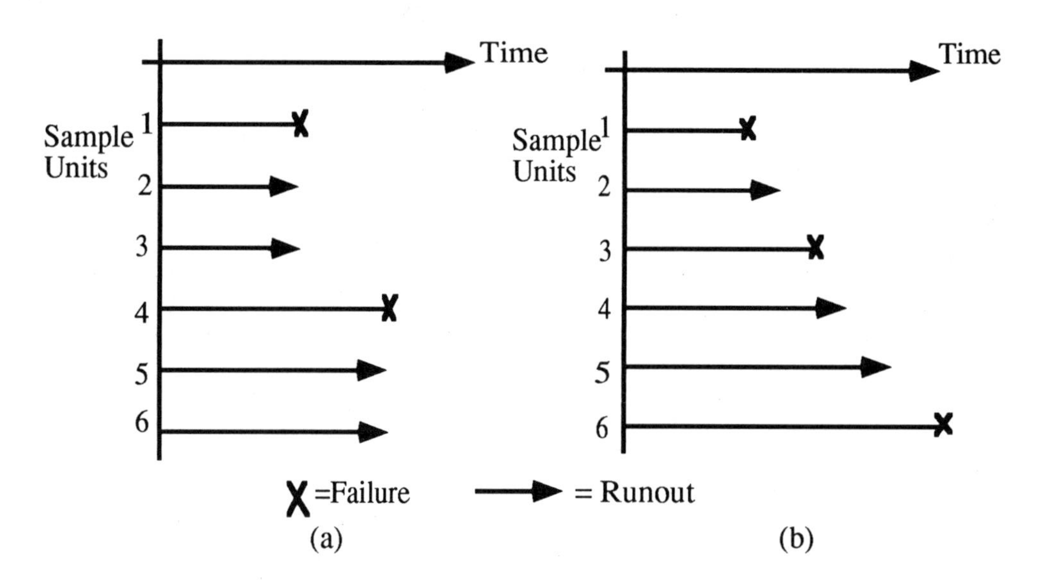

FIGURE 5.4 · *Multicensored data*

Now, back to the questions: How does one determine if a product life is exponential and How does one estimate its parameter?

Reliability data can be analyzed in two ways: using 1) graphical and 2) analytical methods.

Generally graphical methods are simple and easy to learn and use. The analytical methods provide more accurate results although they are a bit involved. Both these methods will be discussed for time-censored, failure-censored, and interval data, but only the graphical method will be given here for the multicensored data.

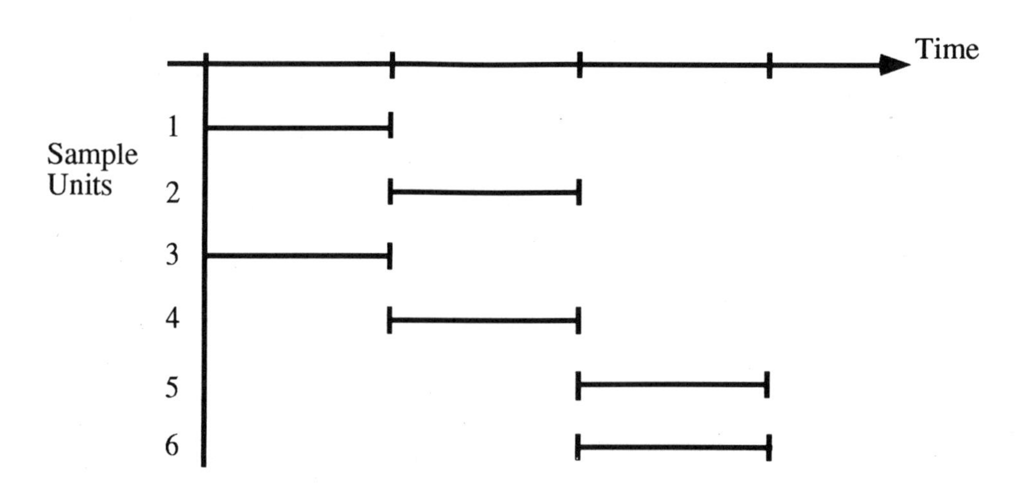

FIGURE 5.5 · *Interval data*

Parameter Estimation for Exponential Distribution

When a product life is suspected of following an exponential distribution, first a *point estimate* for the failure rate is found and then a *goodness of fit* test is used to verify if the distribution is exponential.

As mentioned earlier, a point estimate is an estimate for a population parameter obtained from a single sample (a sample may consist of more than one unit). A point estimate cannot give the true value of the parameter since it has error from the variability it inherits from the variability in the individual measurements themselves. Statistical theory provides methods of quantifying this error and expressing it through the confidence intervals. There will be more on this later, but let's first discuss the methods of calculating a point estimate for the exponential parameter—the failure rate.

The failure rate can be estimated using the formula:

$$\hat{\lambda} = \frac{\text{number of failures}}{\text{total unit test hours}}$$

[The hat notation (^) is used to indicate that it is an estimate for a parameter.]

This formula is good for complete, censored, and interval data. The basic formula takes a little different form for the various types of data when it is written in terms of observations from test.

If n represents the number of units in the sample, r the number of failures in the test, and t_i the number of hours, cycles, etc., at which the ith unit in the sample fails, then for complete data,

$$\hat{\lambda} = \frac{n}{\sum_{i=1}^{n} t_i}$$

for time-censored data censored at time T,

$$\hat{\lambda} = \frac{r}{\sum_{i=1}^{r} t_i + (n - r)T}$$

for failure-censored data censored at rth failure,

$$\hat{\lambda} = \frac{r}{\sum_{i=1}^{r} t_i + (n - r)t_r}$$

In each of these formulas the denominator represents total time units are on test, and the numerator the number of failures for that many units of test time. Although these formulas seem to be the logical way the failure rate should be calculated, it should be recognized that at least for the censored data, the rate at which failures occur can be calculated in other ways. Statisticians would say that these formulas provide the best estimates in that they have the desirable properties unbiasedness and minimum variance.

As noted earlier, since MTTF (or MTBF) $= \frac{1}{\lambda}$, an estimate of λ also provides an estimate for MTTF which is denoted by Θ. The following examples show how failure rate and MTTF are estimated from experimental results.

Example Complete Data

Six fuel pumps were put on test. All failed at the times shown. If the time to failure is known to be exponential, what is the estimate for failure rate and MTTF?

Failure No.	1	2	3	4	5	6
Failure Time (hr)	1000	1500	2200	3700	5600	7900

Solution For complete data:

$$\hat{\lambda} = \frac{n}{\sum_{i=1}^{n} t_i} = 6/21900$$

$$= 0.000274 \text{ failures/hr}$$

$$\hat{\Theta} = \frac{1}{\hat{\lambda}} = 3650 \text{ hrs}$$

Example Time Censored Data

In a life test of five batteries, failures were observed after 10, 30, and 40 hrs. The fourth and fifth batteries were tested for 75 hours without failure, at which time the test was terminated. Find an estimate for λ and Θ if the time to failure is known to be exponential.

Unit No.	1	2	3	4	5
Time (hr)	10	30	40	75+	75+

(The + sign indicates the unit is not a failure but a survivor.)

Solution For time-censored data:

$$\hat{\lambda} = \frac{r}{[\sum_{i=1}^{r} t_i + (n - r)\, T]}$$

$$= \frac{3}{[(10 + 30 + 40) + (2)(75)]}$$

$$= \frac{3}{230} = 0.013 \text{ failures/hr}$$

$$\hat{\theta} = \frac{1}{0.013} = 76.9 \text{ hrs}$$

Example Failure-Censored Data

A sample of eight randomly selected cooling fan motors were observed in their field use. The motors were known to have exponential life and the failure rate was to be calculated after the fifth failure. Estimate the failure rate and MTTF if the failure times were as shown below.

Unit No.	1	2	3	4	5	6	7	8
Time (hr)	400	2000	8000	10,600	18,700	18,700+	18,700+	18,700+

Solution For failure-censored data:

$$\hat{\lambda} = \frac{r}{[\sum_{i=1}^{r} t_i + (n - r)\, t_r]}$$

$$= \frac{5}{[39700 + (3)(18700)]}$$

$$= \frac{5}{95800}$$

$$= 5.22 \times 10^{-5} \text{ failures/hr}$$

$$\hat{\theta} = \frac{1}{\hat{\lambda}} = 19160 \text{ hrs}$$

Example Repairable Units

When the equipment in question is repaired and put in service after a failure and continues to be observed as sample unit, a different situation arises and is illustrated in this example.

A sample of five auto transmissions were identified in the field and observed over a period of time. The hours logged by each transmission and the number of repairs in each were as shown below. If the transmissions have been debugged and so early failures are not possible, and no wearout failures are expected within the test period, what is the estimate for the constant failure rate? What is the estimate for MTBF during the "useful life period" (this is another name for the period of chance failures) of the transmissions?

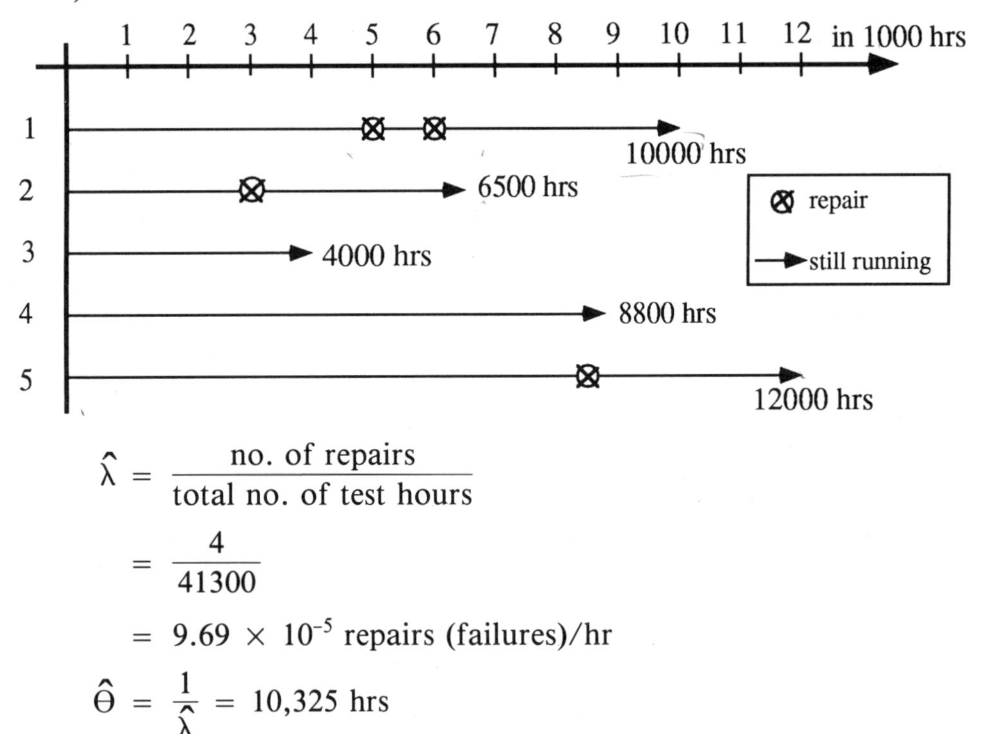

$$\hat{\lambda} = \frac{\text{no. of repairs}}{\text{total no. of test hours}}$$

$$= \frac{4}{41300}$$

$$= 9.69 \times 10^{-5} \text{ repairs (failures)/hr}$$

$$\hat{\theta} = \frac{1}{\hat{\lambda}} = 10{,}325 \text{ hrs}$$

Example

Interval Data

One hundred microswitches were tested for durability using a special test fixture and the numbers failing in different cycle intervals is shown in Table 5.1.

INTVL-CYCLES	NO. FAILR	CUM FAILR	NO. SURV	FAILURE RATE
c = 0	0	0	100	
0<c≤70	45	45	55	[45/70]/100 = 0.0064
70<c≤140	20	65	45	[20/70]/55 = 0.0052
140<c≤210	13	78	22	[13/70]/45 = 0.0041
210<c≤280	7	85	15	[7/70]/22 = 0.0045
280<c≤350	3	88	12	[3/70]/15 = 0.0029
350<c≤420	6	94	6	[6/70]/12 = 0.0071
420<c≤490	3	97	3	[3/70]/6 = 0.0071
490<c≤560	3	100	0	[3/70]/3 = 0.0142
	100			Σ = 0.0515
				avg = 0.0064

TABLE 5.1 · *Failure Data on Microswitches*

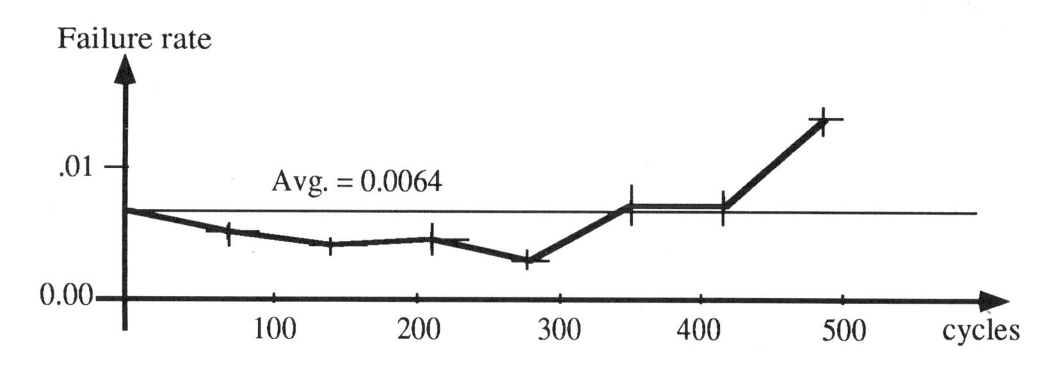

FIGURE 5.6 · *Failure rate function of microswitches*

Solution For this case the failure rate is calculated at the beginning of each interval and is seen to be reasonably constant over the test period (see Figure 5.6). The average value of 0.0064 failures per cycle can be taken as the constant failure rate. The MTTF (in this case it is in fact mean number of cycles to failure) = 156.25

Goodness of Fit Test for Exponential Distribution

A goodness of fit test is used to ascertain if a population has the characteristics of a hypothesized distribution. A set of sample data is taken from the population and the frequency of occurrence in the data in different regions is compared with the expected frequency of occurrence in those regions had the data come from the hypothesized distribution. If there is wide deviation between the expected and the observed frequencies, the hypothesis is rejected. Otherwise there would be no reason to believe that the population in question does not have the hypothesized distribution.

There are two popular goodness of fit (gf) tests: 1) chi-squared gf test and 2) Kolmogorov-Smirnov gf test. These tests could be used to test for any distribution such as normal, exponential, etc. The chi-squared gf test is described in the following steps as it applies to the exponential distribution.

Chi-Squared Goodness of Fit Test—for Exponential Distribution

Step 1: Group the data as would be done for making a histogram. Let A_i be the actual number of observations in each cell, n total number of observations in data, and k number of cells.

Step 2: Find a point estimate $\hat{\lambda}$ for the exponential parameter λ and propose the hypotheses to be tested as:

Null hypothesis H_o: the distribution is $\text{Exp}(\hat{\lambda})$

Alternate hypothesis H_1: the distribution *is not* $\text{Exp}(\hat{\lambda})$

Step 3: Estimate the expected frequency of observations in each cell assuming the population does follow the hypothesized distribution, using the following formula.

Expected number of observations in each cell: $E_i = n[e^{-\hat{\lambda} L_i} - e^{-\hat{\lambda} U_i}]$

where L_i and U_i are the lower and upper limits of each cell, respectively.

Step 4: If any cell has expected number of observations less than five, combine the cells so that each cell has at least five observations. The cell widths need not be all equal.

Step 5: Calculate $(A_i - E_i)^2/E_i$ for each cell and find the total of this quantity for all cells. This quantity represents how much difference there is between the pattern of data and the hypothesized distribution. This quantity is known to have the chi-squared (χ^2) distribution with degrees of freedom (df) equal to (k - 2), where k is the number of cells. Call this χ^2 (obs) for observed value of the χ^2 variable.

Step 6: For a selected value of α, the probability of error that can be allowed, find from the chi-squared table (Table A.3) the critical value of the χ^2 with df $= k - 2$. (Use df equal to k - 1 if the λ was not estimated from the data but a target was given.)

Step 7: If χ^2 (obs) $>$ χ^2(crit), reject H_0. That is, conclude that the data do not come from the hypothesized distribution. Otherwise, do not reject H_0.

Example Using the chi-squared goodness of fit test verify if the data shown in Table 5.1 come from an exponential distribution.

Solution

Step 1: See Table 5.2.

Step 2: $\hat{\lambda} = 0.0064$ as already estimated

H_0: The microswitches have exponential life with $\lambda = 0.0064$

H_1: H_0 not true

Steps 3,4, and 5: See Table 5.2

Steps 6 and 7: Choose $\alpha = 0.05$, χ^2 (0.05, 4) $= 9.49$; χ^2 (obs) is not greater than χ^2 (crit). Do not reject H_0. There is no reason to believe the microswitch life is not exponentially distributed.

CELLS	A_i		E_i			$\frac{(A_i - E_i)^2}{E_i}$
0 < c < 70	45		$100[e^{-.00640 \times 0} - e^{-.0064 \times 70}]$	= 36.1		2.19
70 < c < 140	20		$100[e^{-.0064 \times 70} - e^{-.0064 \times 140}]$	= 23.1		0.42
140 < c < 210	13		100[.408 - .261]	= 14.7		0.20
210 < c < 280	7		100[.261 - .167]	= 9.0		0.44
280 < c < 350	3		100[.167 - .106]	= 6.1		
350 < c < 420	6	9	100[.106 - .068]	= 3.8	9.9	0.08
420 < c < 490	3		100[.068 - .043]	= 2.5		
490 < c < 560	3	6	100[.043 - .028]	= 1.5	6.8	0.09
560 < c	0		100[.028]	= 2.8		
				Total		3.42

TABLE 5.2 · *Example of χ^2 goodness of fit test*

Confidence Interval for λ and MTTF

As stated earlier, when λ is estimated from one set of samples or one test, the point estimate obtained from it cannot be taken to be the true value of λ. The estimate obtained from one set of samples will be different from another set of samples from the same population. In other words, the estimate has variability in it.

Statisticians have studied this variability in the estimates of parameters and have created methods for measuring this variability and the error it causes in the estimates. The confidence interval can be looked upon as a method of expressing the estimated value of a parameter along with a statement of this error.

A confidence interval can be interpreted as follows: Suppose a 95% confidence interval for MTTF is given as [12,800, 18,600] hrs, that means there is a 95% chance that the true value of the MTTF lies in the stated interval; there is a 5 percent chance that it is not in this interval.

Sometimes one-sided confidence bounds are used. If, for example, it is stated that 99% (one-sided) lower bound for MTTF is 6000 hrs, it means the MTTF in question is at least 6000 hrs, and there is a 1% chance the statement is in error.

The confidence statements for λ or Θ (MTTF) are made based on the knowledge of the distributional characteristics of the estimates. The distribution of the estimates $\hat{\lambda}$ and $\hat{\Theta}$ are known to be related to the χ^2 distribution. That is the reason the CIs for λ and Θ are made using the percentiles of the χ^2-distribution.

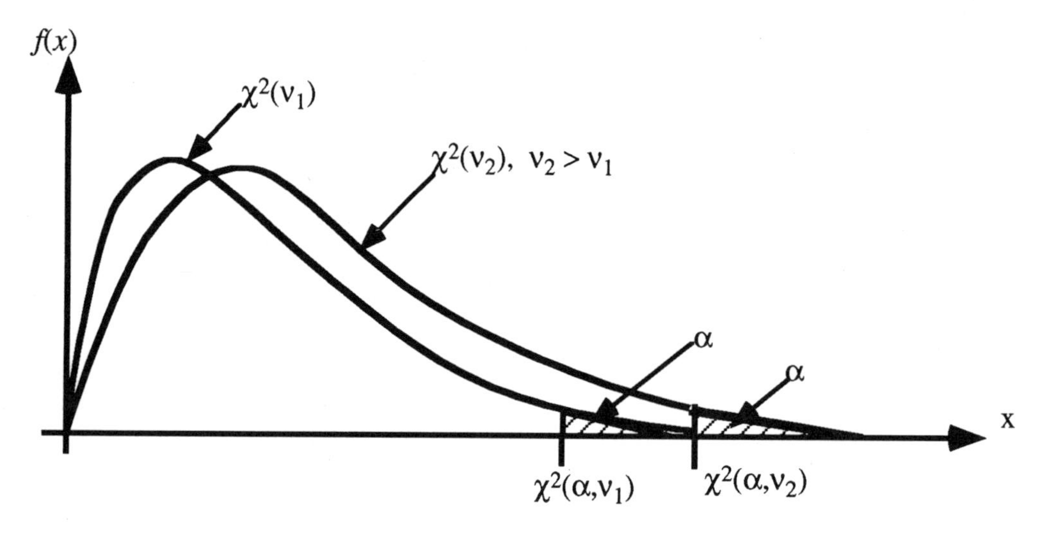

FIGURE 5.7 · *The χ^2 distribution.*

The χ^2 distribution, as noted in Chapter 3, is one of the theoretical distributions used to model random variables. It is a member of the gamma family and has a single parameter called the *degrees of freedom* (df). It is a positive-valued distribution and has the skewed shape as shown in Figure 5.7, which changes with the value of the parameter. The smaller the value for df the closer is the peak to zero.

The percentile values of the χ^2 distribution available in tables such as Table A.3 are $\chi^2(\alpha)$ values such that they cut off α probability at the upper tail of the χ^2 distribution with the stated degrees of freedom.

Confidence Interval for λ from Failure-Censored (Type II) Data

$100(1 - \alpha)\%$ CI for λ is given by:

$$\left[\hat{\lambda}\frac{\chi^2\left(2r, 1-\frac{\alpha}{2}\right)}{2r},\ \hat{\lambda}\frac{\chi^2\left(2r, \frac{\alpha}{2}\right)}{2r}\right]$$

where $\hat{\lambda}$ is the point estimate for λ, r is the number of failures and $\chi^2(2r, \alpha/2)$ is the value of the χ^2 variable with df $2r$ that cuts off $\alpha/2$ probability in the upper tail; and $\chi^2(2r, 1 - \alpha/2)$ is the value that cuts off $(1 - \alpha/2)$ on the upper tail (or $\alpha/2$ at the lower tail) of the distribution. For a method of deriving these results, see Nelson[15].

Once CI for λ is obtained, the CI for MTTF or MTBF can be obtained easily by taking the reciprocals of the limits for λ. However, the reciprocal of the lower confidence limit for λ provides the upper confidence limit for MTTF and vice versa. The procedure is illustrated in the following example.

Example A sample of eight cooling fan motors was observed in field and after five

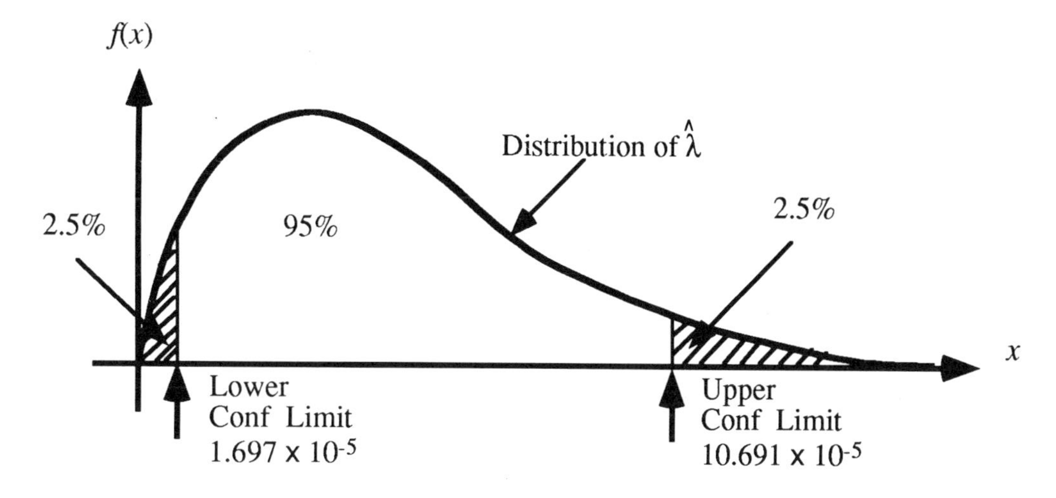

FIGURE 5.8 · *Confidence interval for failure rate of cooling fan motors*

failed, the estimate for λ was obtained as 5.22×10^{-5} per hour. Find a 95% CI for the failure rate and MTTF assuming the life of the motors is exponential.

Solution

$\hat{\lambda} = 5.22 \times 10^{-5}$ per hour

$2r = 2 \times 5 = 10$

$1 - \alpha = 0.95 \quad \rightarrow \alpha = 0.05 \quad \rightarrow \alpha/2 = 0.025 \quad \rightarrow 1 - \alpha/2 = 0.975$

From the χ^2 table:

$\chi^2\,(10, 0.975) = 3.25 \qquad \chi^2(10, 0.025) = 20.48$

95% CI for λ:

$$\left[\frac{(5.22 \times 10^{-5}) \times 3.25}{10}, \frac{(5.22 \times 10^{-5}) \times 20.48}{10}\right]$$

$= [1.697 \times 10^{-5}, 10.691 \times 10^{-5}]$ failures per hr.

This confidence interval is the interval within which the true value of λ is expected to fall 95% of the time. This is illustrated in Figure 5.8. [The distribution of $\hat{\lambda}$ shown in Figure 5.8 is not strictly χ^2, but a positive valued distribution related to the χ^2 distribution.]

It can be seen that the lower limit of this confidence interval provides the 97.5% lower bound, in the sense that the probability the true value of λ is at least this value is 0.975. The upper limit provides a 97.5% upper bound since the probability that the true value is below this quantity is 0.975. As a rule, the 100γ% two-sided confidence limits will give $100(1 + \gamma)/2$% one-sided confidence bounds.

The 95% confidence interval for Θ for the fan motors is given by:

$$\left[\frac{1}{10.691 \times 10^{-5}}, \frac{1}{1.697 \times 10^{-5}}\right] \text{hrs} = [9354, 58928] \text{ hrs}$$

Note that the reciprocal of upper confidence limit for λ gives the lower confidence limit for Θ and vice versa.

Confidence Interval for λ from Time-Censored (Type I) Data:

$100(1 - \alpha)\%$ CI for λ is given by:

$$\left[\hat{\lambda}\frac{\chi^2(2r, 1 - \frac{\alpha}{2})}{2r},\ \hat{\lambda}\frac{\chi^2(2r + 2, \frac{\alpha}{2})}{2r}\right]$$

Note the difference in df for the χ^2 variable for the lower and upper limits.

Example

A production line conveyor broke down 10 times within 200 days since its installation. If the time between failures can be assumed to be exponential, find a 90% confidence interval for λ and MTBF for such production line conveyors.

Solution

This is time-censored data.

$$\hat{\lambda} = \frac{10}{200} = 0.05 \text{ failures/day}$$

$$\hat{\Theta} = \frac{200}{10} = 20 \text{ days}$$

$$2r = 20;\ 1 - \alpha = 0.9 \rightarrow \alpha = 0.1 \rightarrow \alpha/2 = 0.05$$
$$\rightarrow (1 - \alpha/2) = 0.95$$

From the χ^2 tables: $\chi^2(20, 0.95) = 10.85$; $\chi^2(22, 0.05) = 33.92$

$$90\% \text{ CI for } \lambda: \left[\frac{0.05 \times 10.85}{20}, \frac{0.05 \times 33.92}{20}\right]$$

$$= [0.027, 0.085] \text{ failures/hr}$$

$$90\% \text{ CI for } \Theta: \left[\frac{1}{0.085}, \frac{1}{0.027}\right] = [11.76, 37.04] \text{ hrs}$$

Example

Certain engine hose has to meet the specification that calls for an MTTF of at least 100,000 hrs with 95% confidence. Twenty-five randomly selected hoses were put on test and one failed at 12,600 hrs at which time the test was concluded. Does the test result provide evidence that this brand of hoses meets the specification?

Solution

This is failure-censored data. We have to find a lower bound for MTTF. First we will find the upper bound for λ.

$$\hat{\lambda} = \frac{1}{12600 + 24 \times 12600} = 3.17 \times 10^{-6} \text{ failures/hr}$$

$$95\% \text{ upper bound for } \lambda = \hat{\lambda}\,\frac{\chi^2(2,\,0.05)}{2} = \frac{(3.17 \times 10^{-6} \times 5.99)}{2}$$

$$= 9.5 \times 10^{-6} \text{ failures/hr}$$

$$95\% \text{ lower bound for } \Theta = \frac{1}{9.5 \times 10^{-6}} = 105{,}263 \text{ hrs}$$

The 95% lower bound for Θ is larger than the specified minimum. Hence, there is no reason to believe that the hoses do not meet the specification.

Zero Failure Model

When a test has to be stopped with none of the sample units failing within reasonable time, as it happens with products with really long life, the number of failures is 0 and the point estimate for λ will also be 0. An upper bound for λ, and hence the lower bound for MTTF can still be obtained using the following formula:

$$(1 - \alpha)100\% \text{ upper bound for } \lambda\text{: } \frac{\chi^2(2,\,\alpha)}{2nt}$$

where n is the number of units tested and t is the test time for each unit.

Example

Two grinder motors were put on test. Neither of them had failed at 1000 hours when the test was stopped. Does the test provide evidence that MTTF of these motors is not less than 1000 hrs at 90% confidence?

Solution

$$90\% \text{ upper bound for } \lambda = \frac{\chi^2(2,\,0.1)}{2nt} = \frac{4.61}{4000}$$
$$= 1.15 \times 10^{-3} \text{ failures/hr}$$

$$90\% \text{ lower bound for } \Theta = \frac{1}{1.15 \times 10^{-3}} = 868 \text{ hrs}$$

No, the test does not provide evidence that the motors meet the minimum MTTF requirement. The 90% lower bound is below the specified minimum MTTF.

Here is an example where the uninitiated would have made an obvious mistake. The statistical method takes into account the fact that the two units of motor are part of a larger population, there is variability from unit to unit, this variability can be modeled using the exponential distribution and the decision could be made based on the prediction from such a model. Mistakes could happen even with the use of the statistical methods, but the chance of making the mistake is measured and the conclusion is drawn such that this chance of error is kept to a very small amount.

Graphical Methods for Exponential Data

Graphical methods involve the use of specially designed graph papers for different distributions such as exponential and Weibull, which are used to verify if a given set of data come from a hypothesized distribution. Once a particular distribution is seen to fit a data set, then several predictions can be made regarding the life of the population from which the data came.

Graphical methods are of two types: 1) probability plotting and 2) hazard plotting.

Probability plotting involves graphing the cumulative percent failures against time. Hazard plotting involves graphing the cumulative hazard (failure rate) function against time. For probability plotting, the cumulative percent failures or cumulative probabilities of failures usually are computed using what are known as the mean rank or median rank approximations. The method becomes somewhat complicated while analyzing the multicensored data. The hazard plotting method, on the other hand, is the same for all types of data and involves only some simple computations. Both methods should produce exactly the same results. Only the hazard plotting method is described here.

The special graph papers used in the graphical analysis of reliability data are available from several sources. (One of them is TEAM, Box 25, Tamworth, NH 03886.) These graph papers are so designed that the plot of a chosen function, cumulative percent failures, or cumulative hazard function, produces a straight line if the data came from the distribution for which the graph paper has been designed. If a set of data does not produce a straight line on one graph paper, plotting on other papers may be tried until a satisfactory fit is obtained. The graphical method for the exponential distribution is illustrated in the following steps and in the following examples.

Hazard Plot for Exponential Distribution

The method described below can be used for all types of data, i.e., complete, time-censored, failure-censored and multicensored data.

1. Order the n times from the smallest to the largest; both failure times and runout times are included in this ordering.

2. Assign reverse ranks to these times. That is, if there are n time values, the smallest time gets rank n, the next smallest gets rank $(n - 1)$, etc.

3. Calculate the hazard value for each failure (no hazard value is calculated for runout times) as $100/k$ where k is the reverse rank for that failure. These are the observed instantaneous failure rates ($\times$ 100 for convenience) at each of those times.

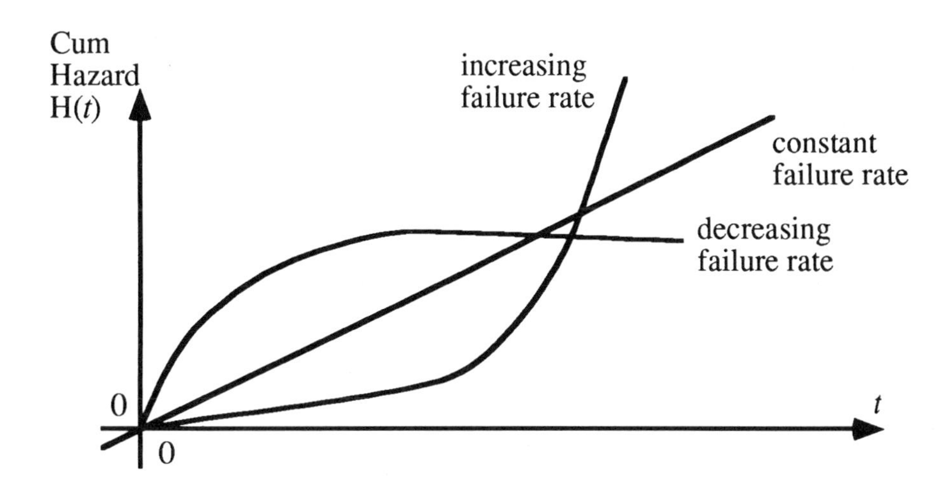

FIGURE 5.9 · *Exponential hazard plots—nature of failure rate*

4. Calculate the cumulative hazard values for each failure. Cumulative hazard values may exceed 100% and have no physical interpretation.

5. Choose the *exponential hazard paper* and plot the cumulative hazard values against each of the failure times. For exponential plots, the scale for the time axis should start at zero.

6. If the plotted points fall in a straight line, the exponential distribution can be taken to provide a good fit for the data. A line is drawn through the plotted points and (0,0). The MTTF can then be read as the time corresponding to cumulative hazard value of 100%. The hazard paper also has a probability scale which gives the cumulative percent failing against time. This scale can be used to read percent failure before a given time or the time to failure of a certain percentage. The latter represents the time before which a certain percentage fail and is denoted by B_{γ}. For example, the B_{10} life represents the time before which 10% of the population would fail.

7. When the data do not plot as a straight line on the exponential paper, plotting them on the hazard paper of another distribution can be attempted. When the plot does not follow a straight line, however, the curvature of the plot gives an indication on the nature of the underlying distribution as shown in Figure 5.9.

Example

Table 5.3 gives the time to failure, in hours, of eight printed circuits. Make a hazard plot and check if the life distribution of the control circuits has an exponential distribution. Estimate MTTF.

Life Hrs	Reverse Rank k	Hazard Value $100/k$	Cum Hazard Value
80	8	12.5	12.5
134	7	14.3	26.8
148	6	16.7	43.5
186	5	20.0	63.5
238	4	25.0	88.5
400	3	33.3	121.8
482	2	50.0	171.8
890	1	100.0	271.8

Table 5.3 · *Hazard Calculation for Control Circuit Life.*

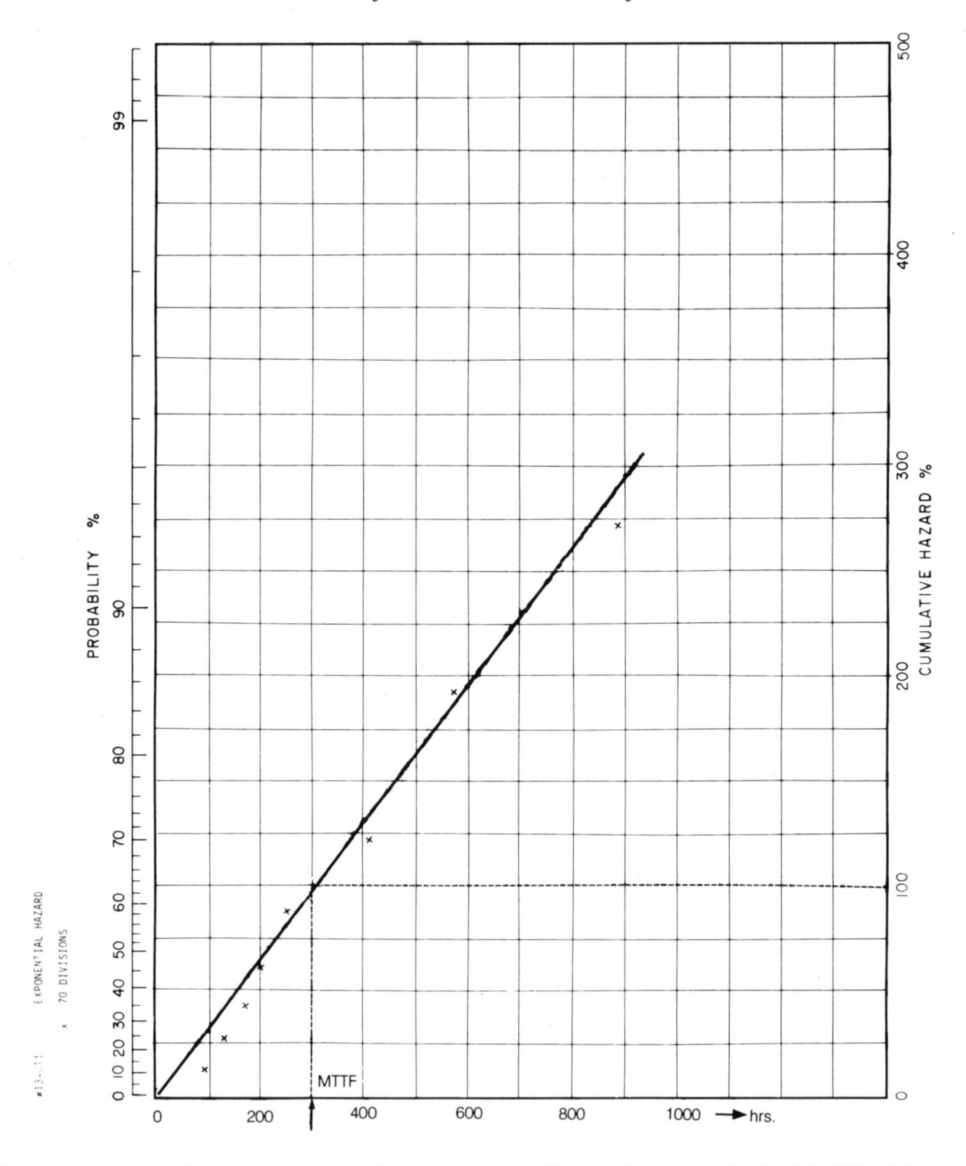

Figure 5.10 · *Hazard plot of printed circuit failures (Data in Table 5.3 TEAM Paper #13-011)*

Solution The graph of the hazard plot is shown in Figure 5.10. The MTTF is read from the graph as 300 hours. It is to be noted that the MTTF corresponds to the 63.2nd percentile as read from the probability scale.

Example Consider the multicensored data on life of crankshaft seals on 20 engines shown in Table 5.4. Make a hazard plot on exponential paper and check the fit.

LIFE MONTHS	REVERSE RANK k	HAZARD VALUE $100/k$	CUM HAZARD VALUE
32	20	5.0	5.0
39	19	5.26	10.26
58	18	5.56	15.82
65+	17	–.–	–.–
66	16	6.25	22.07
70	15	6.67	28.74
75	14	7.14	35.88
75+	13	–.–	–.–
88	12	8.33	44.21
88+	11	–.–	–.–
94+	10	–.–	–.–
102+	9	–.–	–.–
106	8	12.50	56.71
109	7	14.28	70.99
110+	6	–.–	–.–
130	5	20.00	90.99
150+	4	–.–	–.–
155	3	33.33	124.32
185	2	50.00	174.32
210	1	100.00	279.32

(+): runout time

TABLE 5.4 · *Hazard Calculations for Crankshaft Seal Life*

The graph of cumulative hazard function is shown in Figure 5.11. An exponential distribution does not fit the data. The crankshaft seals have increasing failure rate as seen by the curvature of the plotted line.

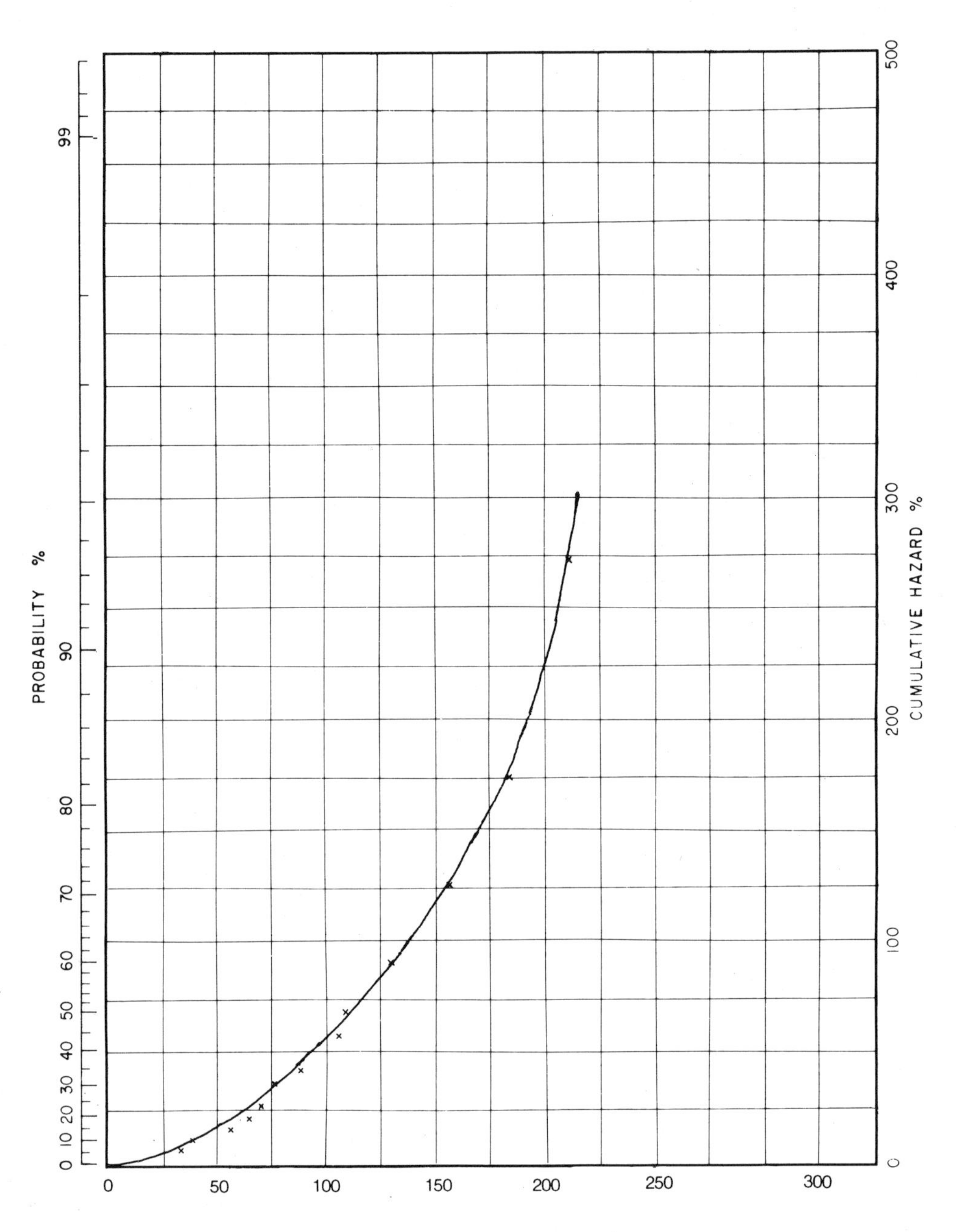

FIGURE 5.11 · *Exponential hazard plot of crankshaft seal life (Data in Table 5.4, TEAM paper #13-011)*

The Weibull Distribution

The exponential would be the first choice when one tries to fit a distribution to a set of life data, because there are more life variables following exponential than any other distribution. When the exponential distribution does not fit a given data, fit for other distributions should be tried. There are several possible distributions one can try such as: normal, log normal, extreme value, gamma, and Weibull. Of all these, the one distribution that seems applicable to several product lives and offers several alternative distributions for consideration is the Weibull distribution.

The Weibull distribution is in fact a family of distributions. The reliability function for an equipment that has Weibull life distribution has the form:

$$R(t) = e^{-\left(\frac{t}{\Theta}\right)^{\beta}}$$

The Weibull distribution has two parameters β and Θ; β is called the *shape parameter* (or *slope*) and Θ is called the *characteristic life*. A more general three-parameter (the third parameter is called the minimum life) Weibull distribution is also available, but the two-parameter model is the one commonly used. Note that when $\beta = 1$ the $R(t)$ assumes the form of the exponential reliability function. The Weibull can be looked upon as a generalization of the exponential, that assumes different shapes for different values of β, as shown in Figure 5.12.

Figure 5.12 also shows the failure rate curves for Weibull distributions with different β values. When $0 < \beta < 1$, the distribution has a decreasing failure rate (DFR). When $\beta = 1$ the distribution is exponential with constant failure rate (CFR) and when $\beta > 1$ it has an increasing failure rate (IFR). Thus, the Weibull distribution can be used to model product lives with a wide range of failure rate behavior. Note that when $\beta = 1$ the shape of the distribution is that of the exponential and when $\beta \cong 3.5$ the shape resembles the normal distribution.

The reliability function for Weibull given earlier, or its complement the cumulative distribution function (CDF) provides the basic definition of the Weibull distribution. All other functions such as the pdf and the failure rate function can be derived from it. These functions of the Weibull are summarized below.

Cumulative distribution function (CDF): $F(t) = 1 - e^{-\left(\frac{t}{\Theta}\right)^{\beta}}$

Reliability function: $R(t) = e^{-\left(\frac{t}{\Theta}\right)^{\beta}}$

Probability density function (pdf): $f(t) = \frac{\beta}{t}\left(\frac{t}{\Theta}\right)^{\beta} e^{-\left(\frac{t}{\Theta}\right)^{\beta}}$

Failure rate function: $h(t) = \frac{\beta}{t}\left(\frac{t}{\Theta}\right)^{\beta}$

These functions can be used, once parameter values are known, for obtaining predictions of reliability, percentiles, and failure rates, analytically. There are procedures for estimating the Weibull parameters from data, using what is known as the maximum likelihood estimation (MLE) method. Computer programs are available to make the estimates for the

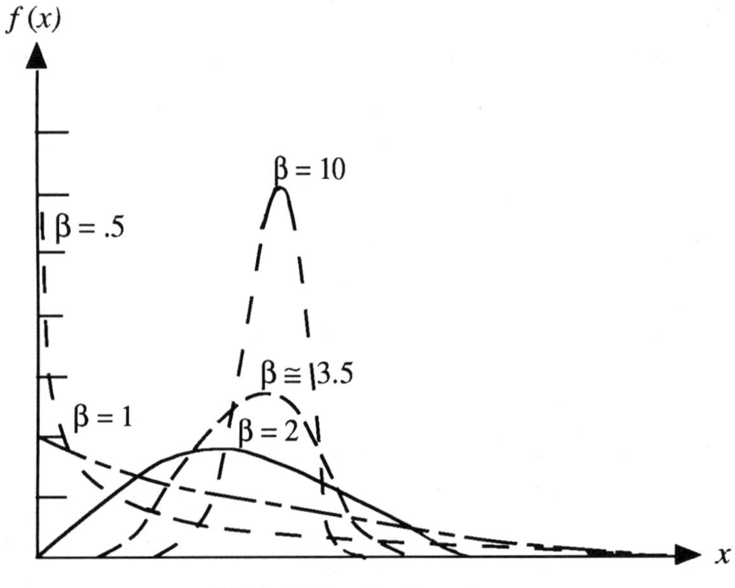

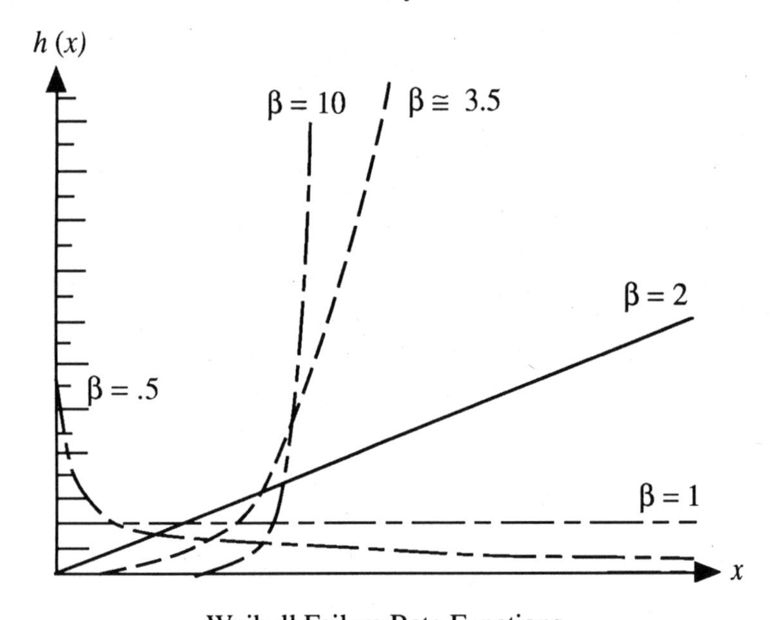

FIGURE 5.12 · *The Weibull distributions and their failure rate function*

Weibull model. The discussion here, however, will be restricted to the graphical methods which are simple, yet informative.

Weibull Hazard Plotting

The computations necessary for hazard plotting for the Weibull is the same as given earlier for the exponential case. The plotting, however, must be done on a paper designed for the Weibull. The hazard plotting procedure for Weibull is illustrated in the example below.

Example Verify, using hazard plot, if the life data of crankshaft seals given in Table 5.4 follow a Weibull distribution. Estimate parameters of the fitted Weibull.

Solution No new calculations are needed since the cumulative hazard function already has been calculated in Table 5.4. The plotting of the cumulative hazard values on a *Weibull hazard* paper is shown in Figure 5.13.

The data plot as a straight line which means the data come from a Weibull population. The parameters can be directly estimated from the graph. The shape parameter β is estimated by drawing a straight line parallel to the fitted line and passing through the point on the graph designated as origin. The intercept made by such a line on the shape parameter scale gives the value of β. (It should be pointed out that the procedure to find β may be different with various brands of graph paper. Usually it is not difficult to figure out how the parameter is to be estimated with a given brand of paper.) The other parameter Θ, called the characteristic life, is estimated as the time value corresponding to 100% cumulative hazard value. It is also the 63.2nd percentile point as seen on the cumulative probability scale. For the example, $\beta = 2.12$ and $\Theta =$ 130 months.

As suspected earlier $\beta = 2.12$ indicates that the seals have an increasing failure rate. It should be pointed out that Θ is *not equal* to MTTF or the average life. It is the 63.2nd percentile of the distribution which means 63.2 percent of the population would have failed by that time. These estimates would be useful if further prediction of reliability measures are to be made analytically or comparison between different designs is to be made based on the parameter values. If β is equal for two designs, a larger value for Θ would mean a longer life.

Most information on a product life, however, can be read directly from the graph. For the example, the following information is read from Figure 5.13.

Percentage failure before 60 months (warranty) = 17.3

Time by which 10% of the population would have failed (B_{10}) = 45 mo

Time by which 1% of the population would have failed (B_1) = 21 mo

Such a plot can be used to compare the life of two similar products, two similar designs, or to track changes in product life.

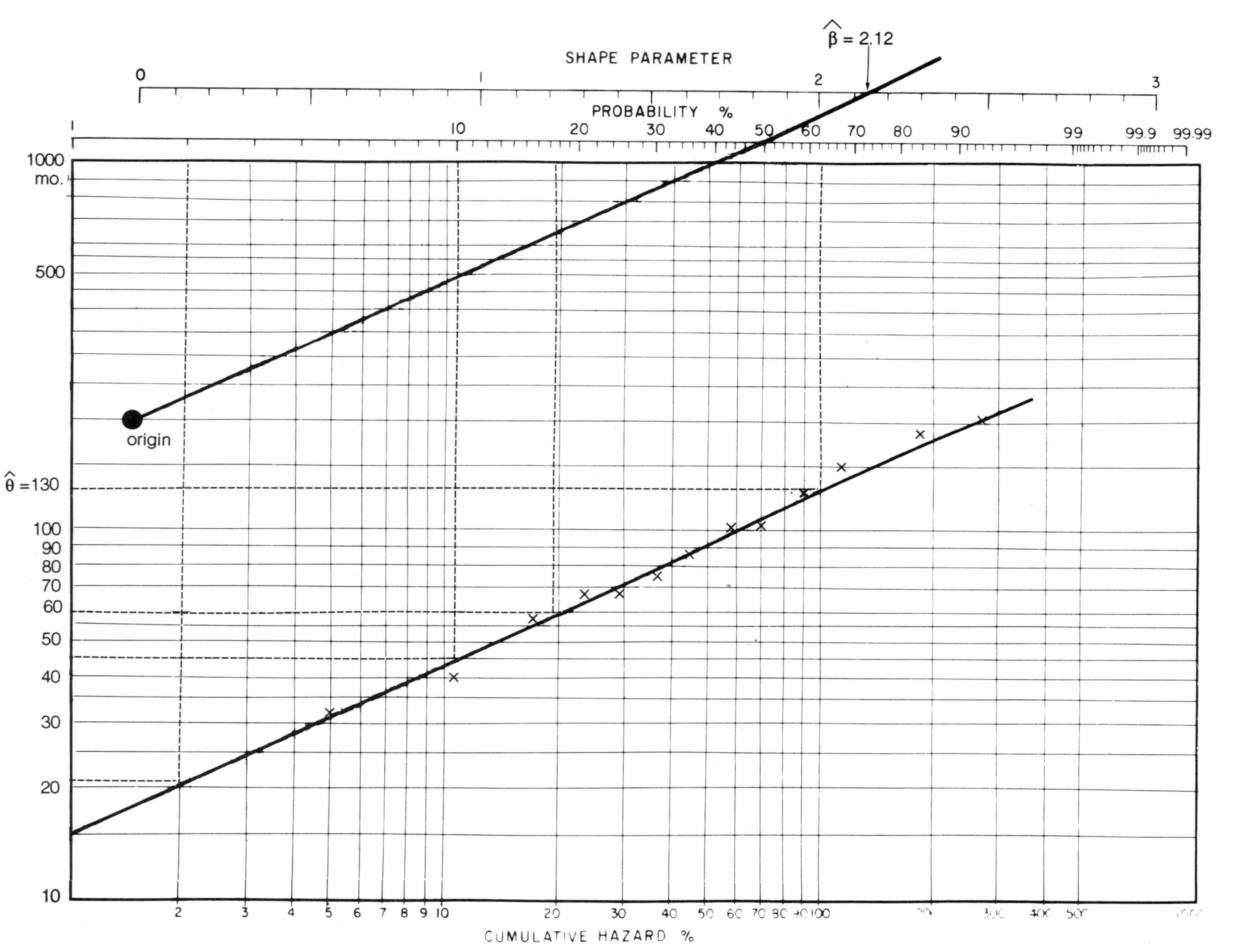

FIGURE 5.13 · *Weibull hazard plot of crankshaft seal life (Data in Table 5.4, TEAM paper #13–181)*

Binomial Distribution

In the discussions above on reliability, it was assumed that failure times of each of the failed units in the sample were known. There are circumstances where the exact failure times of individual units are not known, but only the proportion that failed during a time interval, out of the units put on test is available. Such data can be considered *attribute data* on reliability while the exact time measurements can be considered *variable data*.

As in QC inspection, attribute data from reliability tests provide less information for a given amount of testing compared to variable data. Yet the attribute type of inspections are convenient in certain circumstances as they do not require continuous monitoring of the units on test.

A typical attribute testing would involve putting n units on test and counting the number of failures f after a time period t. Then a point estimate for reliability is given by:

$$R(t) = 1 - \frac{f}{n}$$

From the fact that f is a binomially distributed variable being the number of "successes" out of n independent trials, confidence limit for $R(t)$ can be established using the binomial distribution. To make the confidence limit calculations easy, several nomographs have been created. One set of such nomographs from the book by Lloyd and Lipow[8] is shown in Figures A.1(a)-(d) in the Appendix. The method of computing the confidence limit for reliability and MTTF is shown in the following example.

Example Twenty thermostats were put on a reliability test. After 2000 hours, seven had failed. Set a 95% lower confidence limit for the reliability of the thermostats for 2000 hours. Find a 95% lower confidence limit for the MTTF of the thermostats.

Solution Sample size $n = 20$

Number of failures $f = 7$

Using the graph in Figure A.1(c) for 95% confidence level, the lower confidence limit for the reliability of the thermostats can be read as approximately 0.46.

This lower confidence limit for reliability also can be found using the following formula which uses normal approximation to the binomial: (see Ireson and Coombs[5]–p.19.35)

$$R_L = \frac{n - f - 1}{n + z_\alpha \sqrt{n(f + 1) / (n - f - 2)}}$$

where z_α is the standard normal variate such that $P(Z \geq z_\alpha) = \alpha$

For the example, $z_\alpha = 1.645$ for $\alpha = 0.05$ and

$$R_L = \frac{12}{20 + 1.645\sqrt{(20)(8)/11}} = 0.457$$

Once the confidence limit for reliability is known, the confidence limit for MTTF can be obtained based on an assumption for the failure distribution.

Suppose the life distribution of the thermostats can be assumed to be exponential. Let Θ_L be the lower 95% confidence bound for MTTF. Then,

$$e^{-(2000/\Theta_L)} = 0.45$$

Solving for Θ_L,

$$\Theta_L = 2505 \text{ hrs}$$

Therefore, at 95% confidence level the MTTF of the thermostats is not lower than 2505 hours.

Note that, in this example, through the assumption of exponential distribution for life, the result from attribute data has been used to set the confidence limit for a measurement parameter.

Normal Distribution

The normal distribution has been found to be an appropriate model to describe wearout times of parts and equipment. Figure 5.14 shows a typical life distribution where the earlier part of life has the exponential distribution and the later part follows a normal distribution. The point of time where the wearout begins is of interest while planning replacement or overhaul schedules.

The reliability engineer planning for reliability improvement of products or systems follows different strategies for avoidance of different types of failure modes. The early failures or infant mortalities have to be eliminated by sufficient burn-in, or avoidance of defective parts and subassemblies through the use of appropriate quality control methods. Different design approaches such as selection of stronger material, larger safety factors, or improved environment, etc., are used to minimize failure rates during the exponential phase of life. Replacement, overhaul, or repair are the preventive maintenance methods used to guard against wearout failures. It is in this context that an engineer needs to know when such replacement or repair must be done. The normal distribution is used to estimate the replacement time, since it provides a good model for wearout times just as the exponential provides a model for useful life.

In Figure 5.14, M is called the mean wearout time and T_w is the time

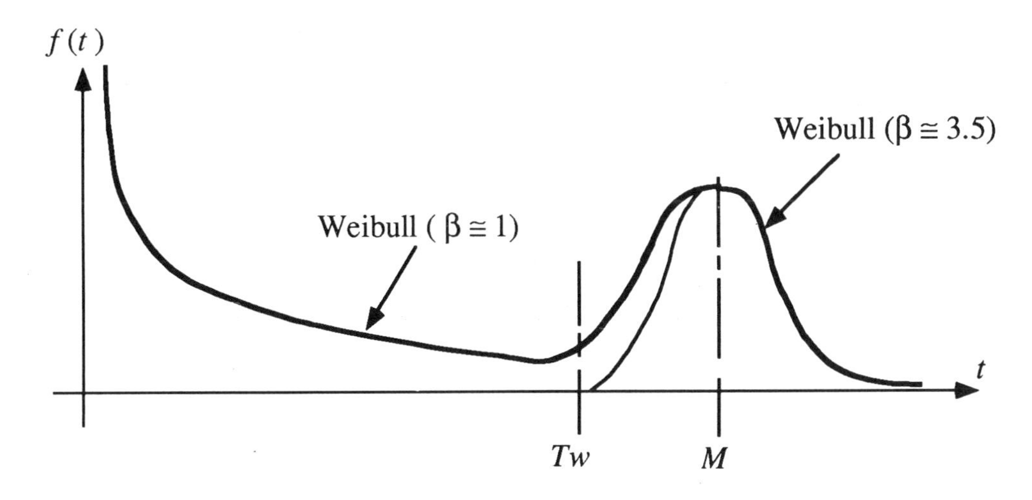

FIGURE 5.14 · *Typical life distribution*

when wearout begins and hence represents time for replacement. An estimate of the average and the standard deviation of wearout times is necessary to determine the replacement time. These estimates must be obtained from test data where the units are tested until they fail due to wearout.

If M and S are estimates for mean and standard deviation of wearout times obtained from a sample of n wearout times, then first a lower bound for the true mean is found:

A 100(1-α)% lower confidence bound for the true mean of wearout times is given by:

$$M_L = M - K_\alpha(S/\sqrt{n})$$

where K_α is the number that cuts off α probability at the upper tail of the standard normal distribution.

Next, T_w is found using the formula:

$$T_w = M_L - K_\gamma S$$

where γ is such that no more than 100γ% of the population will wearout before T_w. K_γ is the number that cuts off γ probability at the upper tail of the standard normal distribution. Figure 5.15 illustrates this procedure.

Values for K_α and K_γ for selected values of α and γ taken from normal table are given below.

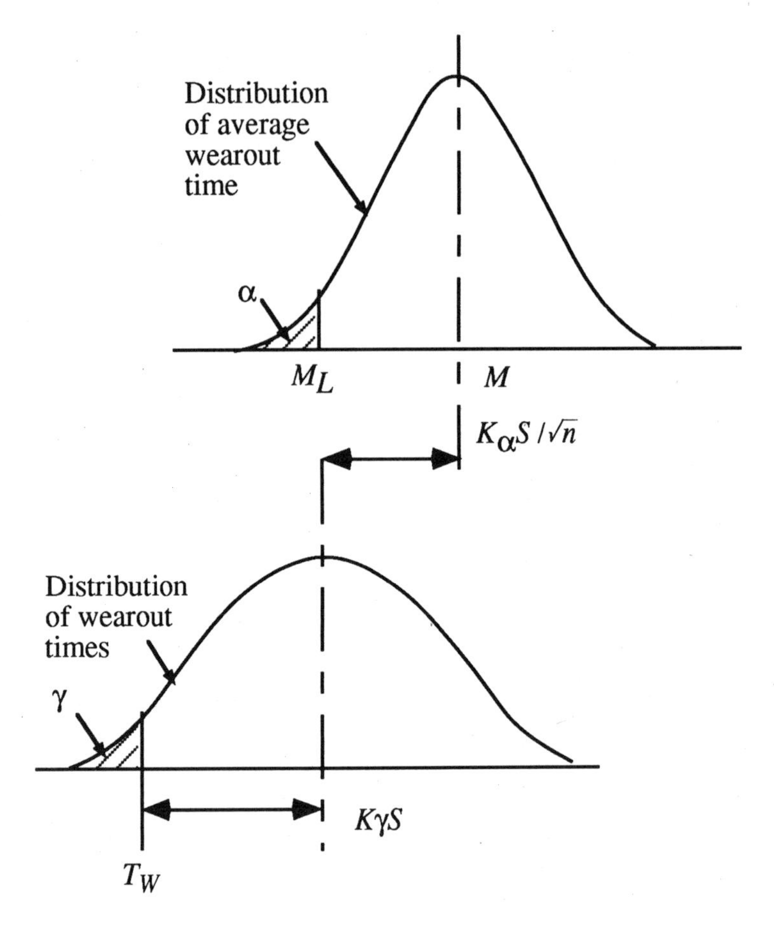

FIGURE 5.15 · *Finding an estimate for replacement time*

α OR γ	ONE-SIDED LEVEL OF CONFIDENCE (PERCENT)	K_α OR K_γ
0.2	80	0.84
0.1	90	1.28
0.05	95	1.64
0.01	99	2.33

Example A test for wearout times for certain type of bushings gave an average = 18,000 hours and standard deviation = 800 hours from a sample of 30 units. Find 90% (one-sided) lower bound for the mean wearout time and find an estimate for the replacement time such that no more than 1% of the bushings will fail due to wearout before they are replaced.

Solution $M_L = M - K_\alpha(S/\sqrt{n})$

$$M = 18000 \qquad S = 800 \qquad \alpha = 1 - .9 = 0.1$$

From tables, $K_\alpha = 1.28$

$M_L = 18000 - 1.28(800/\sqrt{30}) = 17813$ hrs

$T_w = M_L - K_\gamma S$

$\gamma = 0.01 \qquad K_\gamma = 2.33$ (from tables)

$T_w = 17813 - 2.33(800) = 15949$ hrs.

The result can be stated as: If the bushings are replaced at about 16,000 hours 99% of failures due to wearout can be avoided with 90% confidence.

EXERCISE

. . . .

5.1 A brand of home computer has a constant failure rate of 0.006 failures per year.

a) What is the probability that a computer will fail during the second year of operation?

b) What is the probability that a computer will fail after the second year?

c) If the manufacturer sells 2000 units of this computer how many of them will be working after 5 years?

5.2 Using the Chi-square goodness of fit test, check if the data on life of turn signals given in Exercise 4.1 follows an exponential distribution. Use $\alpha = 0.05$.

5.3 The pdf of life (in years) of an electronic component is known to be

$$f(x) = \frac{1}{20} e^{-\frac{t}{20}}, \ t \geq 0$$

What is the reliability of the component for 3 years?

5.4 Twenty turbochargers were put on test after they had been subjected to burn-in to remove early failures. Nine of them failed at the following times: 27, 13, 10, 20, 23, 29, 27, 7 and 14 days. Estimates for λ and μ were made on the 35th day when no more than the 9 units had failed. What are the point estimates for λ and μ? Assume the life follows exponential distribution.

5.5 When the test on the turbochargers in exercise 5.4 was continued, one more failure occurred at the end of 41st day. What are the estimates for λ and μ after the 10th failure?

5.6 There are 32 limit switches in an automatic mold line in a foundry. During 5000 hours of operation of the mold line there were 15 failures of the line due to the limit switches. Each time a switch failed it was replaced. What is the estimate for MTTF of the switches? Assume life is exponential.

5.7 Estimate λ and μ from the following interval data.

TIME INTERVAL (DAYS)	NO. OF FAILURES
$0 \leq t < 3$	21
$3 \leq t < 6$	10
$6 \leq t < 9$	7
$9 \leq t < 12$	9
$12 \leq t < 15$	2
$15 \leq t < 18$	1

5.8 Give a 90% CI for λ and μ of turbochargers in exercise 5.4.

5.9 Give a 90% CI for λ and μ of turbochargers after the 10th failure on the 41st day.

5.10 Give a 95% CI for MTTF of limit switches in exercise 5.6.

5.11 Two high pressure pumps were put on test and one failed at 220 hours and the other had not failed at 380 hours. If the pumps have to meet a minimum MTTF of 300 hours, do the test results show that the pumps meet the specification at 95% confidence?

5.12 In exercise 5.11, if an estimate for λ had to be made at 150 hours when neither of the pumps had failed, what would be the 95% lower bound for μ?

5.13 50 diesel generators were identified as samples of a design, and the following represent data on cooling fan failure times (in hours) on these generators. Verify using hazard plotting, if the fan life is exponentially distributed and if so, estimate the MTTF from graph.

450	460	1150	1150	1560	1600	1660+	1850+	1850+	1850+
1850+	1850+	2030+	2030+	2030+	2070	2070	2080	2200+	3000+
3000+	3000+	3000+	3100	3200+	3450	3750+	3750+	4150+	4150+
4150+	4150+	4300+	4300+	4300+	4300+	4600	4850+	4850+	4850+
4850+	5000+	5000	5000+	6100+	6100	6100+	6100+	6300+	6450+

5.14 The following data are time to failure, in hours, of certain insulator at twice the design voltage. Make a hazard plot of the data on Weibull hazard paper and estimate the parameters.

1, 1, 2, 3, 12, 25, 46, 56, 68, 109, 323, 417

5.15 The time to failure of a bearing seal is known to have the Weibull distribution with $\beta = 2.2$ and $\Theta = 12{,}000$ hours.

a) What is the reliability of the seals for 10,000 hours?

b) What proportion of the seals would have failed before 5,000 hours?

c) What is the failure rate of the seals at 1,000, 5,000 and 10,000 hours?

5.16 The following represent the data from the life test of a toaster switch in number of cycles. Make a hazard plot on Weibull hazard paper and estimate β, Θ and B_{10} life.

32, 39, 58, 65+, 66, 70, 75+, 75+, 88+, 88+, 94+, 102+, 106, 109+, 110, 130+.

5.17 10 out of 36 door locks had failed in a test at the end of 1500 cycles. What is the 90% (one sided) lower confidence bound for the reliability of the door locks for 1500 cycles?

CHAPTER 6

Reliability Apportionment

Apportionment refers to the process of assigning reliability goals to components within a system in order to achieve the reliability goals established for the system. The basic principles that govern such apportionment are discussed herein.

Most systems (e.g., electronic, power, mechanical, electrical, or electromechanical) can be modeled as series, parallel, or standby systems, or a combination thereof. When the components are assembled in one of these configurations, the reliability of the system can be computed from the knowledge of the component reliability.

Series Systems

In a series system the components are connected in such a manner that if any one of the components fail, the entire system fails. Such a system can be schematically represented by a reliability block diagram (RBD) as shown in Figure 6.1.

For a series system, if the assumption can be made that the component failures are independent, the system reliability for time t is given by:

$$R_s(t) = R_1(t) \times R_2(t) \times \ldots \times R_n(t)$$

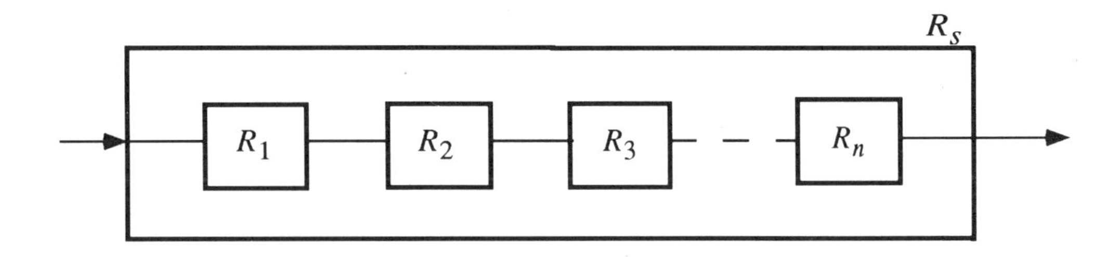

FIGURE 6.1 · *Series system configuration*

where $R_i(t)$ is the reliability of the ith component. This is a direct result of applying the product rule to obtain the probability of joint occurrence of independent events. For the system to function, *all* of the components should function simultaneously.

It can also be shown that, if $h_s(t)$ represents the failure rate of the system and $h_i(t)$ represents the failure rate of the ith component, then:

$$h_s(t) = \sum_{i=1}^{n} h_i(t)$$

In other words, the failure rate of a series system is equal to the sum of the failure rates of its components. This is true no matter what the failure distributions of the components are. Only the components must be mutually independent.

Figure 6.2 depicts this result in a graph, where $h_1(t)$, $h_2(t)$, and $h_3(t)$ represent the failure rates of the components and $h_s(t)$ the failure rate of

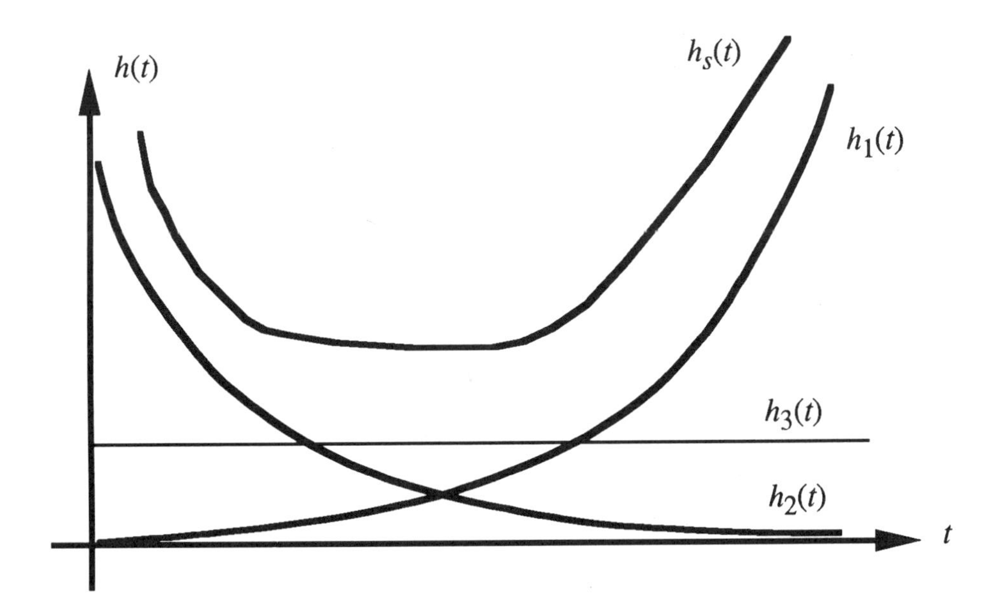

FIGURE 6.2 · *Failure rate of a series system*

the series system made up of the three components. The system failure rate curve which is obtained by summing the individual rates at different t values has the shape of a bathtub. It should come as no surprise that many pieces of complex equipment, which are made up of components having different failure rate characteristics connected in series, have a combined failure rate curve that resembles the bathtub curve.

Example

Find the minimum reliability of the components 3 and 4 in the system shown in Figure 6.3 if the system reliability is to be at least 0.8. The components are independent and components 3 and 4 are identical.

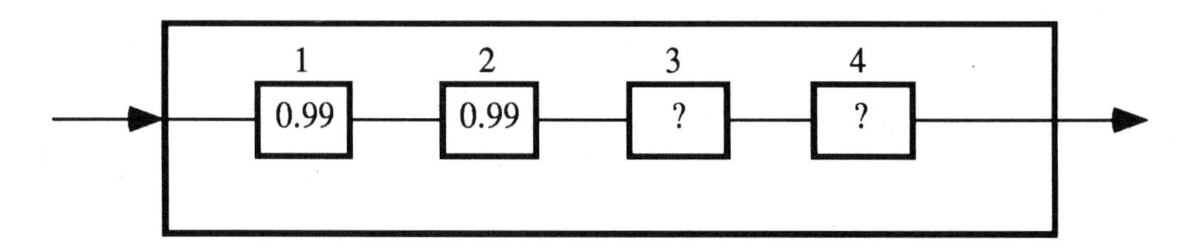

FIGURE 6.3 · *Schematic of a series system*

Solution

Let R be the reliability of 3 and 4

$$0.8 = (0.99)^2 \times R^2$$

$$R^2 = 0.8/(0.99)^2$$

$$R = 0.903$$

Example

An automobile engine fails when any one of its four subsystems fails. The fuel supply (*FS*) system is known to have an exponential life with MTBF = 100,000 hours, the oil supply (*OS*) system has exponential life with MTBF = 170,000 hours, the cooling water (*CW*) system has a Weibull distribution with $\beta = 0.7$ and $\Theta = 120{,}000$ hours, and the electrical system (*ES*) also has a Weibull distribution with $\beta = 0.7$ and $\Theta = 180{,}000$ hours. Determine the reliability of the engine for 20,000 hours.

Solution

$$R_{FS}(20000) = e^{-\frac{t}{\Theta}} = e^{-\frac{20000}{100000}} = 0.819$$

$$R_{OS}(20000) = e^{-\frac{t}{\Theta}} = e^{-\frac{20000}{170000}} = 0.889$$

$$R_{CW}(20000) = e^{-\left(\frac{t}{\Theta}\right)^{\beta}} = e^{-\left(\frac{20000}{120000}\right)^{2.0}} = 0.973$$

$$R_{ES}(20000) = e^{-\left(\frac{t}{\Theta}\right)^{\beta}} = e^{-\left(\frac{20000}{180000}\right)^{0.7}} = 0.807$$

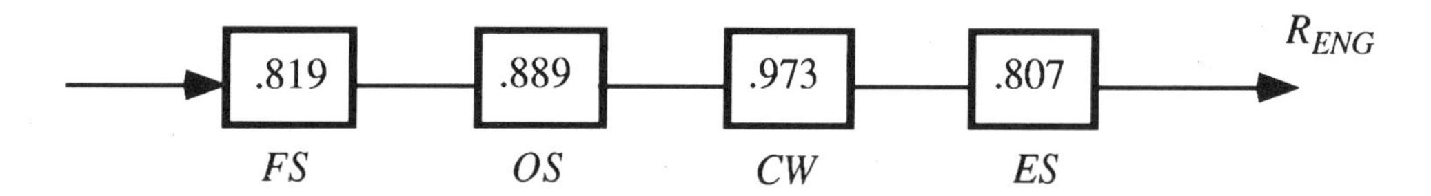

$$R_{ENG} = R_{FS} \times R_{OS} \times R_{CW} \times R_{ES} = 0.572$$

Parallel Systems

In a parallel system, the system fails only when all of the components fail. Such a system's configuration can be represented by the RBD shown in Figure 6.4. In this configuration the n components are live and the system will perform if at least one of the components is working. Such a system also is known as a *redundant system.*

If the components are independent, the reliability of a parallel system for time t is given by

$$R_s(t) = 1 - [(1 - R_1(t)) \times (1 - R_2(t)) \times \ldots \times (1 - R_n(t))]$$

$$= 1 - \prod_{i=1}^{n} (1 - R_i(t))$$

where, R_i is the reliability of the ith component [Π is the product sign such that $\prod_{i=1}^{3} a_i = a_1 \times a_2 \times a_3$]. It is easy to see how this formula will give the reliability of the system. The quantity $\prod_{i=1}^{n} (1 - R_i(t))$ gives the

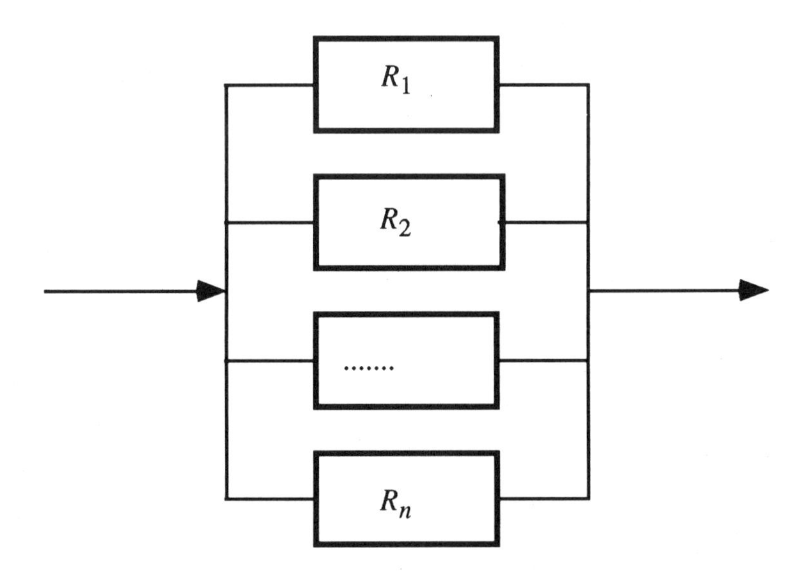

FIGURE 6.4 · *Schematic of a parallel system*

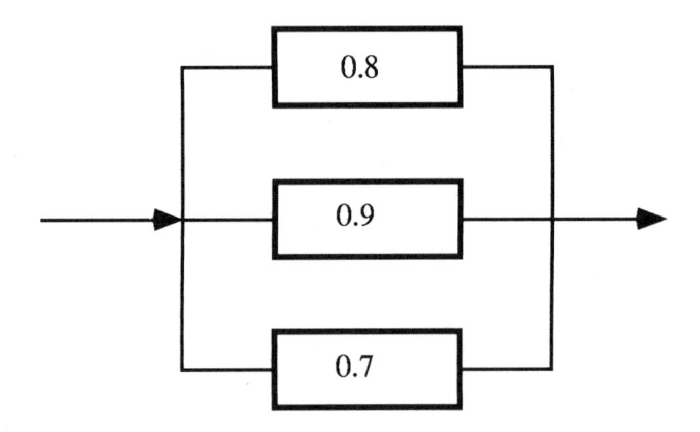

FIGURE 6.5 · *Example of a parallel system*

probability that the components will all fail by time t. Its complement will give the probability that all of them will not fail (i.e., at least one of them will be alive), which is the required reliability.

When all the components are identical and the argument t can be dropped, the above expression for the reliability of a parallel system reduces to a concise form:

$$R_s = 1 - (1 - R)^n,$$

where R is the reliability of each of the identical components for the time under consideration and n is the number of components in the system.

Example Find the reliability of a system that is connected as shown in Figure 6.5 if the component failures are independent.

Solution

$$R_s = 1 - (1 - 0.8)(1 - 0.9)(1 - 0.7) = 0.994$$

Example Find the reliability of the systems shown schematically in Figures 6.6(a) and (b). Assume that the components are mutually independent.

Solution System (a) is a parallel connection of three series systems each having reliability $= 0.8 \times 0.9 \times 0.7 = 0.504$. Therefore the reliability of system (a):

$$R_a = 1 - (1 - 0.504)^3 = 0.878$$

System (b) is a series connection of three parallel systems having reliability: 0.992, 0.999, and 0.973. Therefore, the reliability of system (b):

$$R_b = 0.992 \times 0.999 \times 0.973 = 0.964$$

Both of these systems have the same components, but when they are connected differently the reliability changes. The lesson from this example is that redundancy built in at a lower level of assembly contrib-

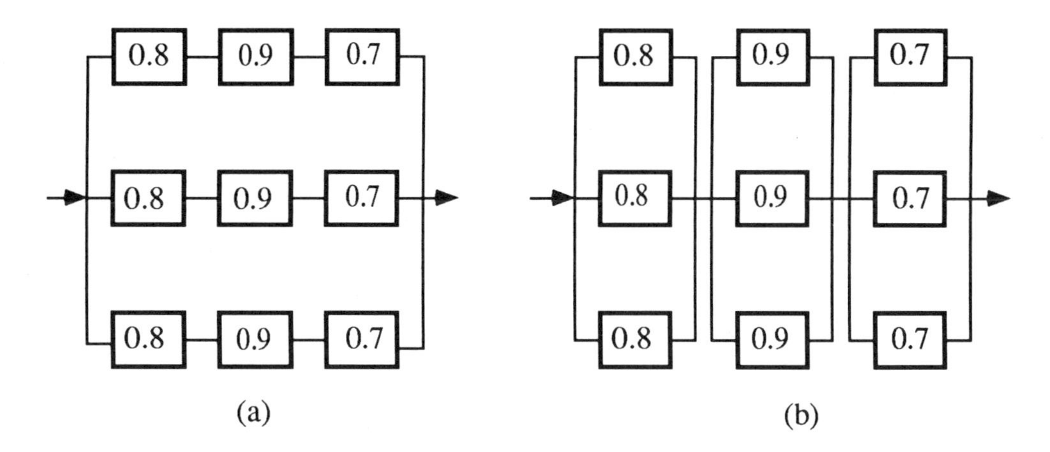

FIGURE 6.6 · *Examples of redundant systems*

utes more toward system reliability than when built in at a higher level of assembly.

Standby Systems

In standby systems the redundant component is activated only after the main component has failed, as opposed to the parallel system in which the redundant component is always live. A switching mechanism is used to kick in the standby unit when the active unit fails. Figure 6.7 shows the standby configuration. For the sake of simplicity the reliability of the switch is assumed to be 1.0.

With the standby configuration, system life is given by the sum of the lives of the individual components. The reliability of a standby system cannot be obtained by application of simple probability rules. Unlike the results for series and parallel systems, the system reliability is dependent on the nature of the life distribution of the components.

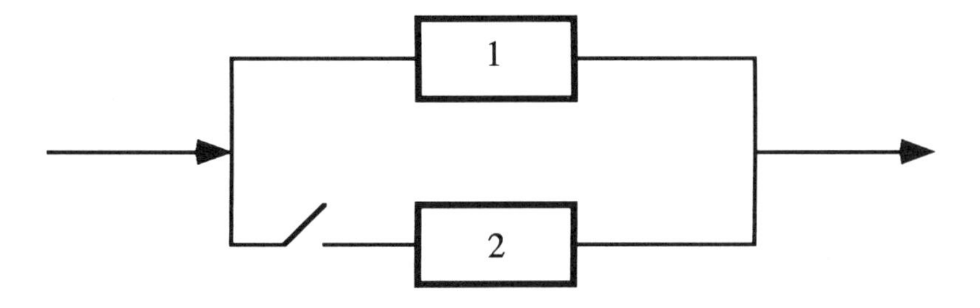

FIGURE 6.7 · *Standby system configuration*

When the component life distribution is exponential with failure rate λ, the reliability of the standby system with r identical, independent components is given by

$$R(t) = e^{-\lambda t} \sum_{j=0}^{r-1} \frac{(\lambda t)^j}{j\,!}$$

This expression for $R(t)$ can be recognized as a summation of terms of the Poisson distribution with average $= \lambda t$, the average number of failures in time t. This comes from the fact that the sum of independent exponential random variables has the gamma distribution and the CDF of the gamma can be obtained from the CDF of the Poisson.

Example Three feed pumps in a power plant are connected in a standby mode and the power plant will have to be shut down if all feed pumps fail. If the failure time of a feed pump has exponential distribution with MTBF = 28,000 hours, what is the probability a power plant failure will not be caused by the feed pump failure in 6000 hours? Assume the switching mechanisms have reliability = 1.0.

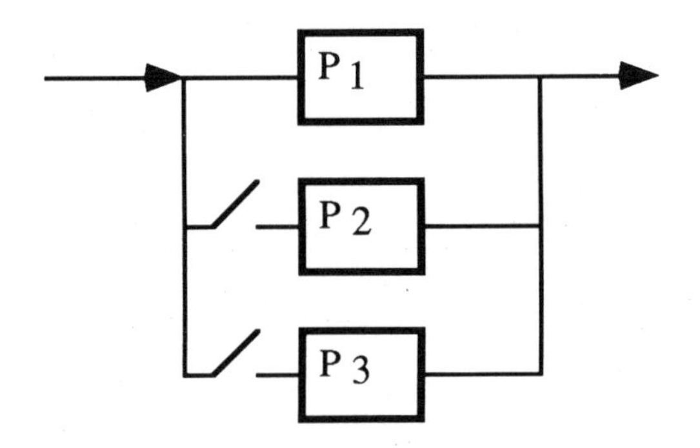

Solution

$$\lambda = \frac{1}{28000} \qquad \lambda t = \frac{6000}{28000} = 0.214 \qquad r = 3$$

$$R(t) = \sum_{j=0}^{2} \frac{e^{-0.214}\,(0.214)^j}{j\,!}$$

$$= e^{-0.214}\left[\frac{0.214^0}{0\,!} + \frac{0.214^1}{1\,!} + \frac{0.214^2}{2\,!}\right]$$

$$= e^{-0.214}\,[1 + 0.214 + 0.0229]$$

$$= 0.9986$$

Standby configuration with the same set of components will result in larger reliability than the ordinary parallel configuration that has all the components operating all the time.

r Out of n-Systems (r/n Systems)

The system consists of n identical, independent components in parallel. The system will work if at least r out of n components are working. If R is the reliability of the components for time t, then the reliability of the system for t is given by:

$$R_s = \sum_{j=r}^{n} \binom{n}{j} R^j (1 - R)^{n-j}$$

This binomial summation gives the probability of r or more of the n components working, which is the probability of the system functioning.

Example A train is pulled by four engines working in tandem. The time between failure of locomotive engines is Weibull distributed with $\beta = 0.8$ and $\Theta = 15{,}000$ hours. At least three engines must be working for the train to be pulled. What is the probability the train will be pulled for at least 15,000 hours without a failure?

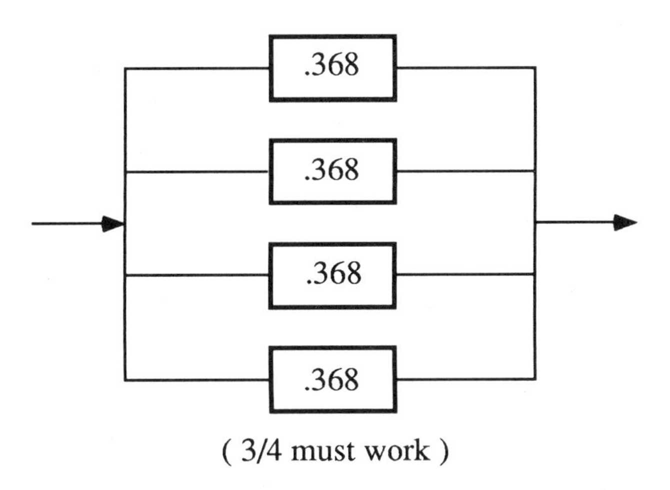

(3/4 must work)

Solution Reliability of one engine for 15,000 hrs:

$$e^{-\left(\frac{15000}{15000}\right)^{0.8}} = 0.368$$

Reliability of the system:

$$R_s = \sum_{j=3}^{4} \binom{4}{j} (0.368)^j (0.632)^{4-j}$$

$$= \binom{4}{3} (0.368)^3(0.632)^1 + \binom{4}{4} (0.368)^4 (0.632)^0$$

$$= 4 \times (0.368)^3(0.632)^1 + 1 \times (0.368)^4 (1)$$

$$= 0.144$$

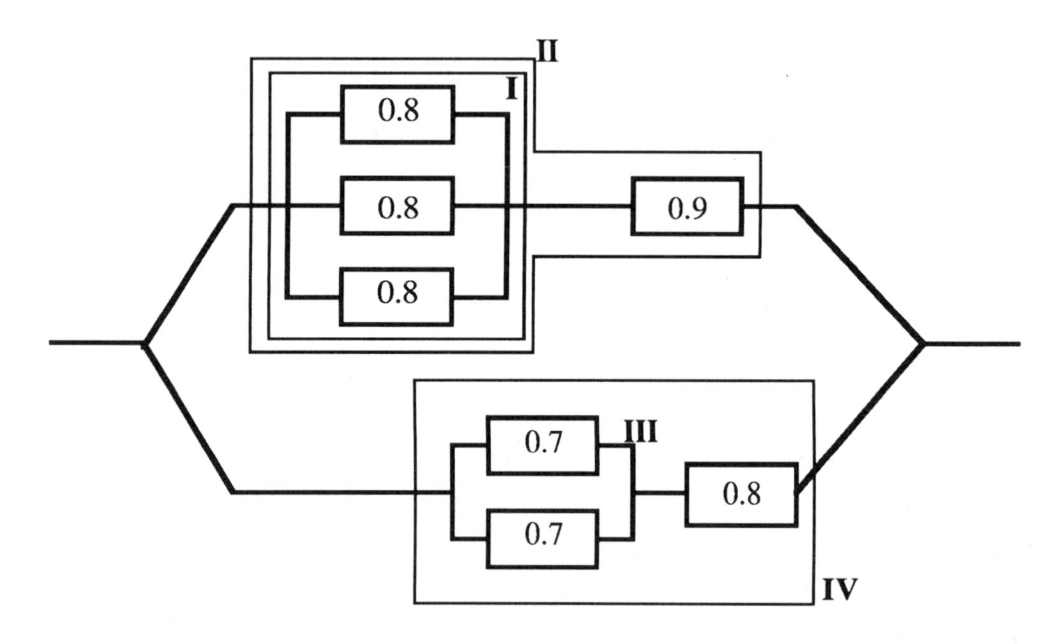

FIGURE 6.8 · *RBD of a complex system*

Complex Systems

Many systems can be represented as a combination of series and parallel systems. Once the system is modeled through a reliability block diagram (RBD), the system reliability can be computed by evaluating reliability of subsystems in a bottom-up manner. Some examples are given to illustrate the methods of analyzing such systems.

Example Evaluate the reliability of the system shown in the RBD in Figure 6.8.

Solution

$$R_{I} = 1 - (0.2)^3 = 0.992$$

$$R_{II} = 0.992 \times 0.9 = 0.893$$

$$R_{III} = 1 - (0.3)^2 = 0.91$$

$$R_{IV} = 0.91 \times 0.8 = 0.728$$

$$R_S = 1 - (1 - R_{II})(1 - R_{IV})$$

$$= 1 - (1 - 0.893)(1 - 0.728) = 0.971$$

Example (Linked System)

The RBD shown in Figure 6.9 is an example of a linked system and there are some special methods for analyzing such a system. One method is illustrated in the following two examples.

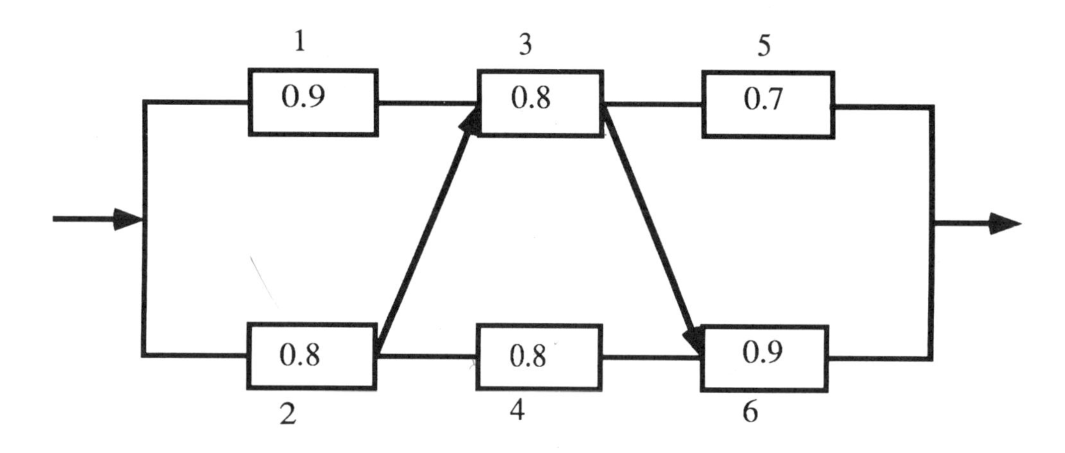

FIGURE 6.9 · *RBD of a linked system*

Example Evaluate the reliability of the system shown in Figure 6.9.

Solution The special method lies in selecting a critical component and finding the conditional reliability of the system with and without the critical component working. Then the theorem on total probability is used to obtain the system's reliability.

For this example, component No. 3 is chosen as the critical component. The system can function with or without it and in each case the system resolves into a simpler system that is easy to analyze.

According to the theorem on total probability, reliability of the system:

$$\begin{aligned} R_s &= P(\text{Syst works} \mid \text{comp 3 is working}) \times P(\text{comp 3 works}) + \\ &\quad (\text{Syst works} \mid \text{comp 3 is not working}) \times P(\text{comp 3 fails}) \\ &= P(\text{Syst works} \mid \text{3 works})R_3 + \\ &\quad P(\text{Syst works} \mid \text{3 not working})(1-R_3) \end{aligned}$$

When 3 works, it does not matter whether or not 4 is working. The system can then be represented by the RBD in Figure 6.10(a). When 3 does not work the system is represented by the RBD in 6.10(b).

With reference to Figure 6.10(a):

$$\begin{aligned} P(\text{Syst works} \mid \text{3 is working}) &= [1 - (0.1)(0.2)][1 - (0.3)(0.1)] \\ &= 0.98 \times 0.97 = 0.9506 \end{aligned}$$

Similarly from Figure 6.10(b):

$$P(\text{Syst works} \mid \text{3 is out}) = 0.8 \times 0.8 \times 0.9 = 0.576$$

$$\begin{aligned} R_s &= P(\text{Syst works} \mid \text{3 is working})R_3 + \\ &\quad P(\text{Syst works} \mid \text{3 is out})\ (1 - R_3) \\ &= (0.9506)(0.8) + (0.576)(0.2) = 0.8757 \end{aligned}$$

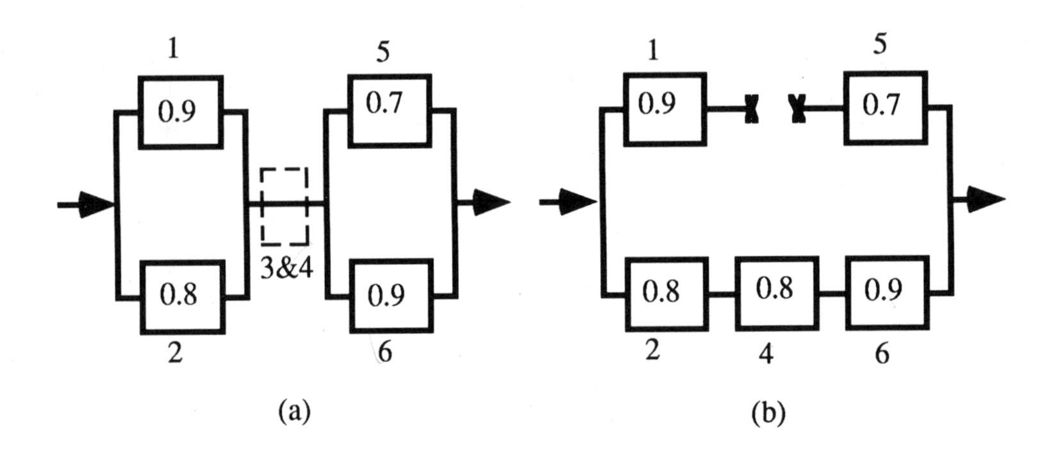

FIGURE 6.10 · *Breakdown of the linked system in Figure 6.9*

Example Find the reliability of the system shown in Figure 6.11

Solution The critical component is E.

$$R_s = P(\text{Syst works} \mid E \text{ is working})R_E + P(\text{Syst works} \mid E \text{ is out})(1 - R_E)$$

With reference to Figures 6.12(a) and (b):

$$P(\text{Syst works} \mid E \text{ is in})$$

$$= [1 - (1 - .95)(1 - .95)][1 - (1 - .99 \times .99)(1 - .99)]$$

$$= 0.9975 \times 0.9998 = 0.9973.$$

$$\text{P(Syst works} \mid \text{E not in}) = 0.95 \times .99 \times .99 = 0.931$$

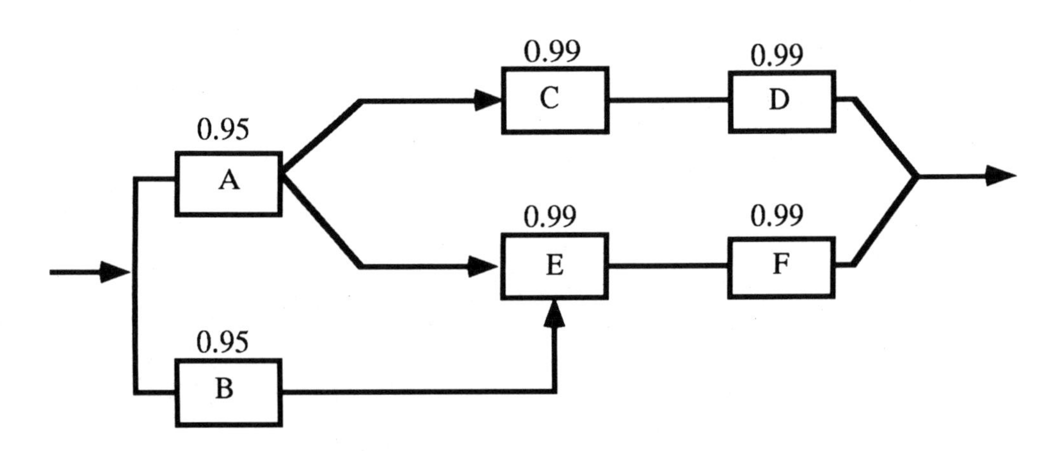

FIGURE 6.11 · *RBD of a linked system*

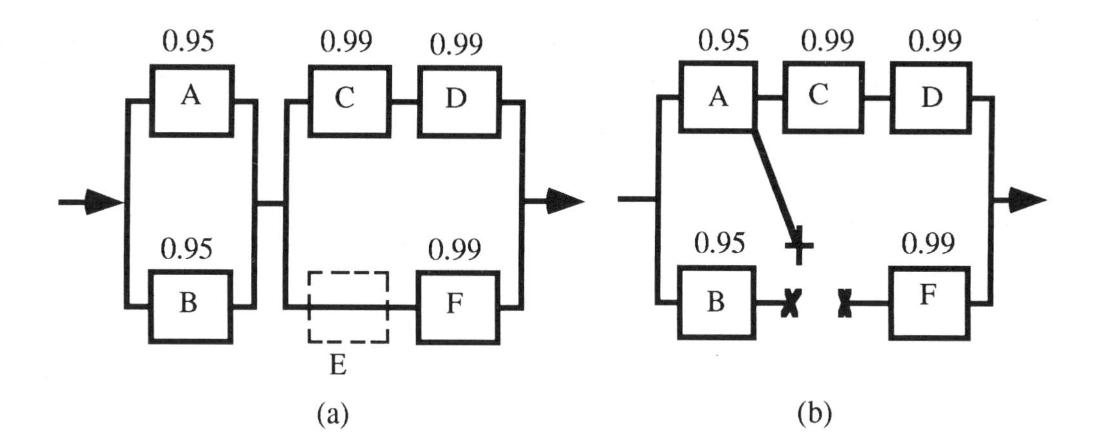

FIGURE 6.12 · *Breakdown of system in Figure 6.11*

$$R_s = P(\text{Syst works} \mid E \text{ in})\, R_E + P(\text{Syst works} \mid E \text{ not in})\,(1 - R_E)$$
$$= 0.9973 \times 0.99 + 0.931 \times 0.01 = 0.9966$$

There are many systems that are more complex than those shown above. There are methods such as reliability networks and computer simulation that can be used to analyze those systems. These are not covered here as they are beyond the scope of this text.

The models discussed above would enable computation of system reliability from component reliability. When the reliability engineer has to determine the distribution of component reliability to achieve a system reliability goal, however, several factors such as economics, time constraint, weight restrictions, and manufacturability enter into the equation. Reliability apportionment is an iterative process to match capabilities with requirements within economic feasibility. The above models will only help in verifying if a chosen distribution of part reliability would achieve the system reliability goal.

EXERCISE

6.1 For a gas furnace to work without failure, the fan, the burner, and the control mechanism should function without failure. If the reliability of the fan is 0.96, that of the burner is 0.98, and that of the control mechanism is 0.7 for 5 years, what is the reliability of the whole furnace for 5 years? Assume the three components fail independent of each other.

6.2 There are 3 safety valves on a boiler drum. At least two of them must be functional in order to relieve any pressure surge. If each valve has 0.70 probability of working when needed, what is the probability that the drum will be safe from pressure surges.

6.3 Four components are connected in parallel in the standby mode with perfect switches, and only one component is needed for the system to function. If the component life is exponential with $\lambda = 0.003$ failures per hour, what is the reliability of the system for 1500 hours?

6.4 Determine the reliability of the system that is shown in the RBD below.

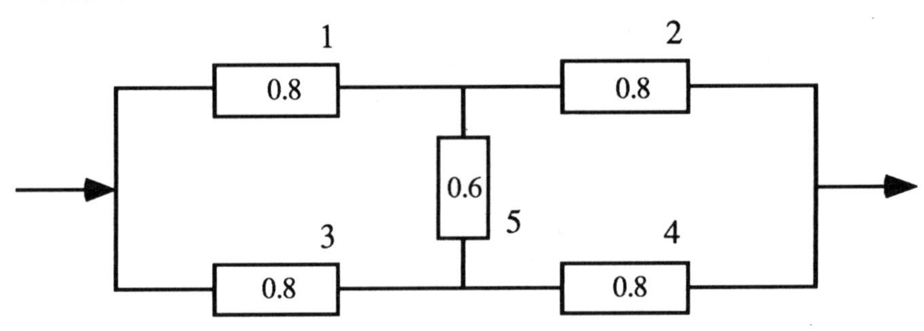

PART III

Reliability Evaluation and Implementation

CHAPTER 7

Reliability Testing

Just as inspection and performance testing is done to evaluate a product's conformance to specification in its physical characteristics and performance requirements, life testing is done to evaluate the reliability of a product over time. Reliability testing can be considered quality appraisal over time.

Plans for reliability testing must be made with extra care since testing for reliability is expensive due to long time periods involved, special fixtures needed, and destruction of at least some of the test units. Plans must take into account the objectives for which testing is done, the levels of risk acceptable, and the constraints such as time available and number of failures that can be allowed.

One of the good sources of reliability test plans is the MIL-STD-781D, "Reliability Testing for Engineering Development, Qualification, and Production," and the accompanying MIL-HDBK-781, "Reliability Test Methods, Plans, and Environments for Engineering Development, Qualification, and Production." These documents provide a collection of ready-made plans suitable to meet the needs of several testing situations and the discussion in this chapter and the next centers around these two documents. The objective of this chapter is to discuss, without the electronic and defense industry lingo, the plans available from them and explain the fundamentals behind the plans where appropriate. The premise is that although these plans had been created to meet the defense needs, there is a wealth of infomation in these that can be of use to many other industries.

The MIL-STD-781D recognizes four different circumstances where test plans will be required:

1. Reliability development/growth testing (RDGT)

This type of testing is done when the current reliability of an equipment/system has to be evaluated, goals for improved reliability established, and test-analyze-and-fix (TAAF) procedure is to be followed until the goals are achieved or exceeded. This approach is used at the development stage of a product when design alternatives are explored and changes are incorporated to enhance the product's reliability. The analysis methods recognize the fact that with each change in design, the life distribution of the product will be different.

2. Reliability qualification testing (RQT)

This is used when a product has been built according to a design and it should be verified if the design would indeed provide the equipment of desired reliability. As a result of this testing designs are approved for production. This also is known as *reliability demonstration* or *design approval* testing.

3. Production reliability acceptance testing (PRAT)

When products are produced from an approved design there is the possibility that production limitations or deficiencies could cause decrease in reliability from the approved design. This type of testing is used where assurance is needed that production deficiencies do not cause loss in reliability.

4. Environmental stress screening (ESS)

Testing is done here under environmental conditions more severe than the normal ambient in which the equipment is expected to perform. Such increased stress will reveal early failures caused by poor material or poor production method. This type of screening ensures that early failures will be eliminated in units before they are shipped to the customer. These tests also are performed on units before they are subjected to qualification and acceptance testing in order that the early failures do not mask the true reliability.

Table 7.1 lists the types of plans suitable for meeting different objectives described above. The remainder of this chapter is devoted to describing the different test plans listed in Table 7.1.

Duane Plot Method

The Duane plot method is a graphical technique used for monitoring growth in reliability during the product development stage. This is a quick and simple method which provides a pictorial presentation of the changes occurring in measured reliability.

	Objective for Testing	Type of Test to Be Used
1.	Reliability development/growth	Duane plot method AMSAA method*
2.	Reliability qualification	Probability ratio sequential test Fixed-duration test
3.	Production reliability acceptance	MTBF assurance test Probability ratio sequential test Fixed-duration test All equipment test
4.	Environmental stress screening	Computed time interval method* Graphical method Standard ESS*

*Methods not discussed in detail in this book.

Table 7.1 · *Reliability Test Plan Selection Based on Test Objectives*

Steps in Making the Duane Plot

Step 1: Select a representative sample

The military standard does not have a recommendation on the number of units (sample size) to be tested. Keeping in mind that this type of testing is done at the design and development stage when production has not begun, the number of available units for testing may be limited. (This testing, however, can be done anytime when improvement in reliability is considered necessary.) The testing could be done on one, two, three, or 10 units; the larger the sample size, the shorter will be the test duration, since the total test time logged by all test units is of interest. The sample size also depends on the type and complexity of the equipment tested; small items such as fuel injectors can be tested with large sample size whereas sample size for large equipment such as truck engines necessarily must be small.

Step 2: Run the test

The sample units are run under the environmental conditions similar to those in which they would normally operate. The performance of the equipment on the selected parameters (such as bhp, fuel consumption, temperature rise, etc.) must be monitored. Nonperformance of the equipment against any parameter will constitute a failure. The total length of test duration is chosen, again, based on practical considerations. The test should continue until at least one failure occurs. Prior knowledge of the equipment's longevity will help in predetermining the total hours (miles, cycles, etc.) of test. The test duration can be modified based on results from the test until reliability goals are accomplished.

Step 3: Data collection and analysis

As the test progresses, the failures along with their times of occurrence are recorded on a failure reporting form. Each time a failure occurs, analysis has to be made to detect the root cause of the failure and appropriate design changes should be incorporated in *all* sample units. The test should be continued in such TAAF mode for the selected test duration. When the test is continued in this fashion, the time between failures will continue to become longer, indicating growth in reliability. If such growth in reliability does not become evident, the test duration will have to be increased.

From the test records, the cumulative number of failures $N(t)$ at certain chosen periods of testing t are recorded, and for each time period $t/N(t)$ is computed. On a log-log paper the values of $t/N(t)$ (which are cumulative MTBFs at time t) are plotted against t. If the plotted points follow a straight line, the Duane model can be considered valid for modeling reliability growth for the equipment in question. A straight line is drawn through the plotted points giving most weight to the most recent points since they contain information from all previous points. Projection for a period beyond the test period can be made assuming the TAAF sequence will be followed in the projected period with the same effectiveness. The Duane model has been found, in practice, to be applicable to reliability growth of many electromechanical equipment.

Example

(The examples in this chapter are simple modifications of those given in the MIL-HDBK-781.)

Three units of a fuel injection system were tested simultaneously until a total of 1000 hours of operation was accumulated. As failures occurred, appropriate design modifications were introduced on all three units. The cumulative number of failures after selected periods of testing were as given below. Verify if the Duane model is appropriate for the growth in reliability of the fuel injectors. Estimate the MTBF achieved at the end of the test period and estimate the MTBF that can be achieved if this development test is continued up to 2000 hours.

t (IN HUNDRED HRS)	$N(t)$ (CUM NO. OF FAILURES)	$t/N(t)$ (CUM MTBF)
1.0	3	0.333
2.0	6	0.333
5.0	3	0.385
8.0	18	0.444
10.0	22	0.454

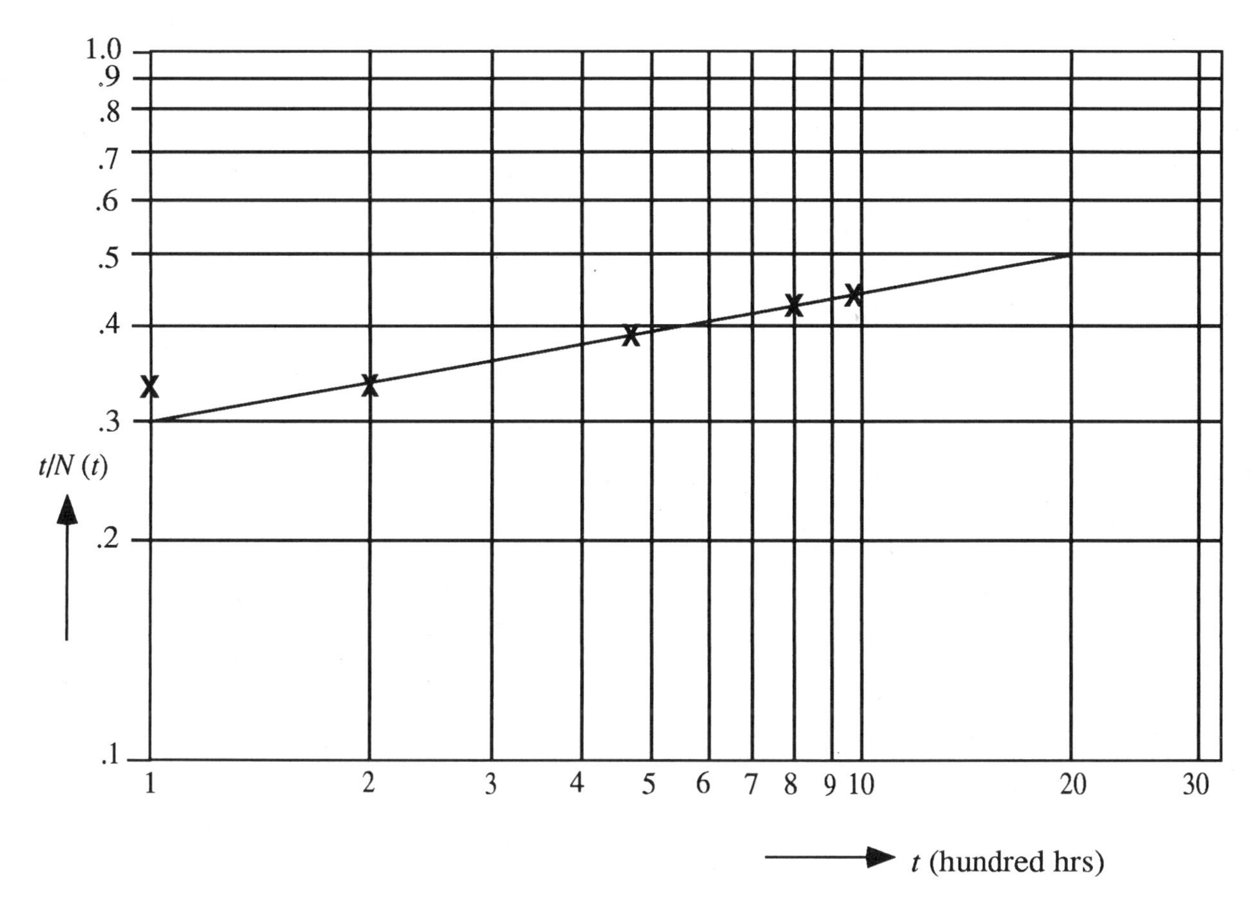

FIGURE 7.1 · *Duane plot of reliability growth of fuel injectors*

The values of t and $t/N(t)$ are shown plotted on a log-log paper in Figure 7.1. The points fall on a reasonably straight line. The progress in reliability seems to fit the Duane model. From the fitted line the current (or instantaneous) MTBF at the end of the test period can be estimated.

The $t/N(t)$ that was plotted on the graph represents the cumulative MTBF at time t, in that it includes all the historical life information up to time t. The value of MTBF of the equipment achieved at the end of the growth testing, i.e., the current (or instantaneous) MTBF at the end of test is given by:

$$\Theta(t) = \frac{t^m}{K(1 - m)}$$

where m is the slope of the Duane plot and K is a constant equal to the reciprocal of the ordinate of the Duane plot at $t = 1$.

For the example:

Ordinate at t = 1: 0.30

$$K = \frac{1}{0.30} = 3.33$$

The slope of the plot can be measured from the graph or can be computed from two sets of values for t and $N(t)$ as

$$m = \frac{ln(0.45) - ln(0.30)}{ln(10) - ln(1)}$$

$$= 0.176$$

Therefore, MTBF at 1000 hours (or 10 hundred hours):

$$\Theta(10) = \frac{10^{0.176}}{3.33(1 - 0.176)} = 0.5465 \text{ hundred hrs}$$

$$= 54.65 \text{ hrs}$$

similarly,

$$\Theta(20) = \frac{20^{0.176}}{3.33(1 - 0.176)} = 0.6175 \text{ hundred hrs}$$

$$= 61.75 \text{ hrs}$$

Note that the cumulative MTBF represented by $t/(N(t)$ at 1000 hrs is smaller than the instantaneous MTBF. The cumulative MTBF has been weighted down by smaller MTBFs in the earlier part of testing.

Principles Behind the Duane Plot Method

The method, originated by J. J. Duane[3] and later refined and modified by others, proposes that the growth in reliability (measured in MTBF) of equipment during a TAAF mode of development can be modeled as:

$$\frac{t}{N(t)} = \frac{t^m}{K}$$

where $t/N(t)$ is the cumulative MTBF (being equal to cumulative units of testing time divided by the total number of failures observed during that time), and K and m are constants.

Therefore,

$$ln\frac{t}{N(t)} = m\, ln(t) - ln(K)$$

This means if $t/N(t)$ is plotted against t on a log-log graph paper, the plot will be a straight line with a slope = m. The constant K can be evaluated as reciprocal of the ordinate at $t = 1$. [Note: $ln(t) = 0$ when $t=1$ and $ln(1/k) = -ln(k)$]

From the above growth model:

$$N(t) = Kt^{1-m}$$

The rate of failure at t

$$= \frac{d}{dt} N(t)$$

$$= K(1 - m)t^{-m}$$

MTBF at t:

$$\Theta(t) = \frac{1}{(\text{failure rate at } t)}$$

i.e.,

$$\Theta(t) = \frac{t^m}{K(1 - m)}$$

Thus, from the plot of the cumulative MTBF instantaneous MTBF at any time can be obtained. MTBF at a future time can also be projected assuming the growth effort will be continued with the same effectiveness.

Apart from the Duane plot method the MIL-HDBK-781 recommends what is called the AMSAA (Army Material Systems Analysis Agency) method. This method assumes that the times between successive failure can be modeled as the intensity function of a nonhomogeneous Poisson process. The method also provides means of computing confidence interval for the chosen reliability measure, which the Duane method does not give. The AMSAA is not discussed here as the simpler Duane method is considered adequate to meet most needs. There also is a separate document MIL-HDBK-189, "Reliability Growth Management," which provides guidelines on managing a reliability growth effort.

Probability Ratio Sequential Tests (PRST)

The PRST (or the sequential test) plans are based on the assumption that *the underlying distribution of time-between-failure is exponential*. These plans can be used for design qualification as well as production acceptance. A set of standard sequential test plans is available in the military handbook which would avoid computing the design for each occasion. These test plans along with their operating characteristics are described in more detail below. First, some definitions are necessary.

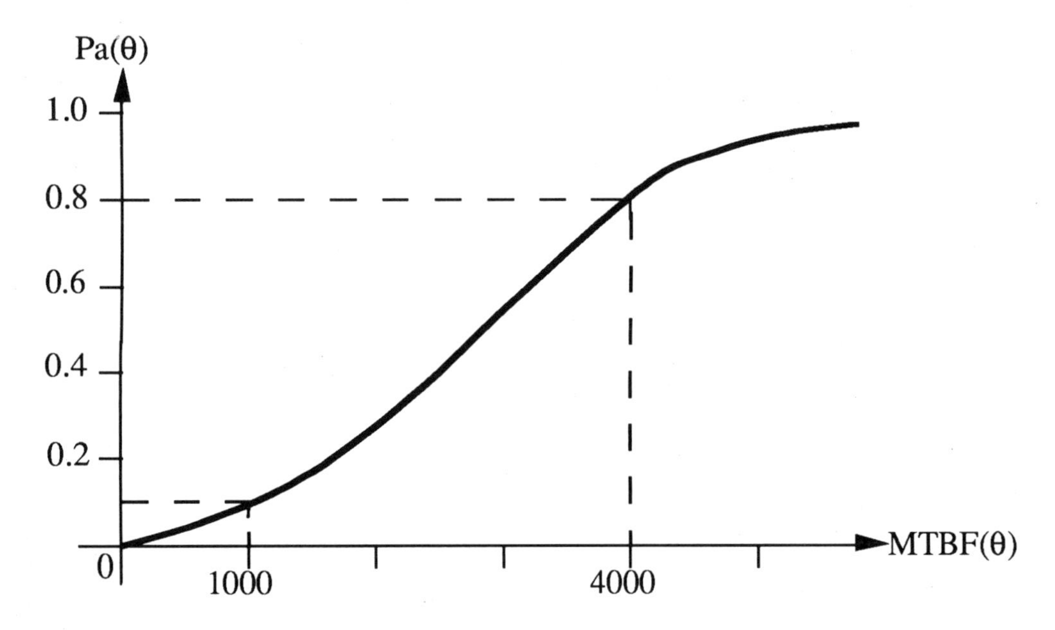

FIGURE 7.2 · *Example of an OC curve of a reliability test plan*

Operating Characteristic Curve (OC Curve)

The OC curve of a test plan shows the relationship between the true reliability (measured in MTBF) of the equipment tested and the probability of acceptance by the plan.

For example, the test plan whose OC curve is shown in Figure 7.2 will accept an equipment that has true MTBF = 1000 hours with probability 0.1, and will accept an equipment with true MTBF = 4000 hrs with probability 0.8. The OC curve enables assessment of the risks involved in accepting equipment with poor (low) MTBF and rejecting equipment with good (high) MTBF. Also, when acceptable risks can be defined, knowledge of OCs of available plans will enable selection of the test plan that would most nearly guarantee the specified risks. The following notations are used in defining the risks:

Lower Test MTBF (Θ_1)–the value of MTBF such that if the equipment has MTBF equal to or smaller than this value, it should be rejected with high probability. This can be called the rejectable MTBF.

Upper Test MTBF (Θ_0)–the value of MTBF such that equipment with MTBF equal to this value or larger should be accepted with high probability. This can be called the acceptable MTBF.

Discrimination Ratio (*d*)–the ratio Θ_0/Θ_1. It is used as a parameter to designate standard test plans. For a given set of risks, smaller values of d will require longer test periods; that is, finer discrimination requires longer testing time.

Producer's risk (α)–the probability of rejecting an equipment with true MTBF equal to the acceptable MTBF (Θ_0).

Consumer's risk (β)–the probability of accepting equipment with true MTBF equal to the rejectable MTBF (Θ_1).

Theoretical Background for Sequential Tests

If the failure time of an equipment has exponential distribution with MTBF of Θ, the probability of obtaining r failures in time t is

$$P(r) = \left(\frac{t}{\Theta}\right)^r \left(\frac{e^{-t/\Theta}}{r\,!}\right)$$

based on the result that if the time between occurrence of events is exponential, the number of occurrences in a time period has Poisson distribution.

If $P_1(r)$ is this probability when MTBF $= \Theta_1$, and $P_0(r)$ the probability when MTBF $= \Theta_0$, respectively, then

$$P_1(r) = \left(\frac{t}{\Theta_1}\right)^r \left(\frac{e^{-t/\Theta_1}}{r\,!}\right)$$

$$P_0(r) = \left(\frac{t}{\Theta_0}\right)^r \left(\frac{e^{-t/\Theta_0}}{r\,!}\right)$$

The ratio $P(r) = P_1(r)/P_0(r)$ is called the probability ratio. If this ratio is quite small indicating P_0 is much larger than P_1, it would indicate the fact that the MTBF is equal to Θ_0 rather than Θ_1; that is, the equipment has acceptable MTBF. On the other hand if the ratio is quite large indicating P_1 is much larger than P_0, it will indicate that the MTBF is equal to Θ_1, rather than Θ_0; that is, the equipment has MTBF not acceptable. The following rules derived from the specified risks are used to determine when the ratio is sufficiently small to result in accept-decision and when it is sufficiently large to warrant reject-decision.

Rules for sequential ratio test:

a) Accept and stop testing when

$$P(r) < B = \frac{\beta}{1 - \alpha}$$

b) Reject and stop testing when

$$P(r) > A = \frac{(1 - \beta)(1 + d)}{2\alpha d} = \frac{1 - \beta}{\alpha} \cdot \frac{1 + d}{2d}$$

c) Continue testing if

$$B \leq P(r) \leq A$$

If a graph is plotted with accumulated time t on the x axis and accumulated number of failures r on the y axis, the above rules of sequential testing can be translated into two parallel straight lines ($a + bt$) and ($c + bt$) where:

$$a = \frac{ln\ B}{ln(\Theta_0/\Theta_1)}$$

$$c = \frac{ln\ A}{ln(\Theta_0/\Theta_1)}$$

$$b = \frac{1/\Theta_1 - 1/\Theta_0}{ln(\Theta_0/\Theta_1)}$$

These lines can be predrawn on a graph and the test can be continued if the (r, t) plot falls between the lines. The test will be stopped with accept decision when a plot falls below the lower parallel line, and with reject decision when a plot falls above the upper parallel line. Figure 7.3 is an example of the graphical representation of the decision rules.

Example

Design a sequential test plan for

$$\alpha = 0.10 \qquad \beta = 0.10 \qquad \Theta_1 = 100 \text{ hrs} \qquad \Theta_0 = 200 \text{ hrs}$$

Determine the accept-reject criteria, and determine the slope and intercepts of the accept-reject lines for graphical plot.

Solution

$$\text{Discrimination ratio} = d = \frac{200}{100} = 2$$

$$A = \frac{(d+1)(1-\beta)}{2d\alpha} = \frac{(2+1)(0.9)}{(2)(2)(0.1)} = 6.75$$

$$B = \frac{\beta}{1 - \alpha} = \frac{0.1}{1 - 0.1} = 0.111$$

$$a = \frac{ln\ B}{ln(\Theta_0/\Theta_1)} = \frac{ln\ 0.111}{ln\ 2} = \frac{-2.198}{0.693} = -3.17$$

$$b = \frac{(1/\Theta_1 - 1/\Theta_0)}{ln(\Theta_0/\Theta_1)} = \frac{0.01 - 0.005}{ln\ 2} = 0.00721$$

$$c = \frac{ln\ A}{ln(\Theta_0/\Theta_1)} = \frac{ln\ 6.75}{ln\ 2} = \frac{1.910}{0.693} = 2.76$$

Figure 7.3 shows the plot of the test plan.

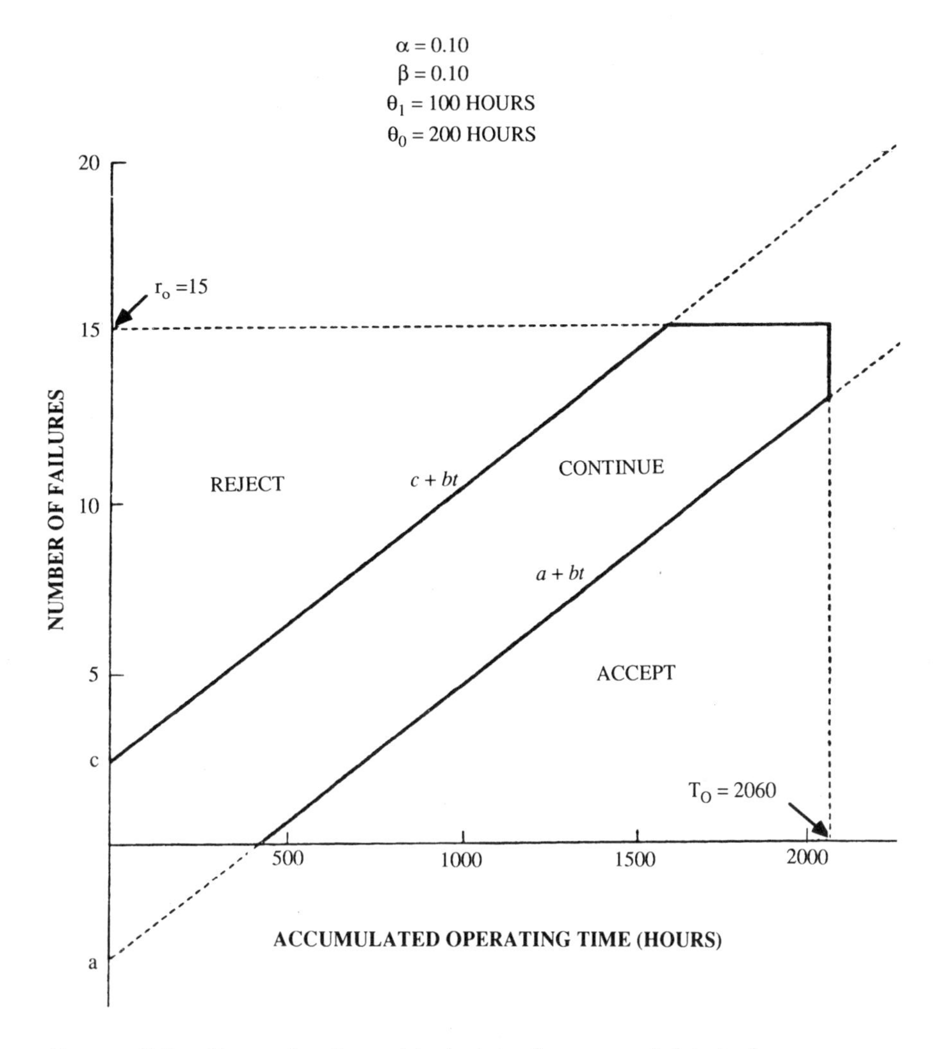

FIGURE 7.3 · *Example of graphical plot of a sequential test plan*

Truncation

Truncation in sequential tests is necessary in order to avoid the uncertainty regarding the length of test time. Truncation can be based on number of failures or test time.

For truncation based on failures, the number of failures r is given by the smallest integer that can satisfy the inequality

$$\frac{\chi^2_{(1-\alpha),2r}}{\chi^2_{\beta,2r}} \geq \frac{\Theta_1}{\Theta_0}$$

where $\chi^2_{(1-\alpha),2r}$ and $\chi^2_{\beta,2r}$ are the values of the χ^2 variable with $2r$ degree of freedom, that cut off $(1 - \alpha)$ and β probabilities, respectively, on the upper tail. The chi-square table must be searched for the value of r that satisfies the above inequality, given Θ_1 and Θ_0. If the resulting r is a fraction, it should be rounded upward to the next integer.

Once the value for r is known, say r_0, then the truncation time T_0 is given by

$$T_0 = \frac{\Theta_0 \chi^2_{(1-\alpha),2r_0}}{2}$$

For the above example:

Needed—the smallest value of r that satisfies:

$$\frac{\chi^2_{0.9,2r}}{\chi^2_{0.1,2r}} \geq 0.5$$

We look in the chi-square table under both 0.9 and 0.1 simulteneously and compute the above ratio for each degree of freedom until we reach the row where the above ratio equals or exceeds 0.5. For this example $2r = 30 \rightarrow r_0 = 15$

$$T_0 = \frac{\Theta_0 \chi^2_{(1-\alpha),2r_0}}{2} = \frac{(200)(20.6)}{2} = 2060 \text{ hrs}$$

Therefore, the test should be truncated at 15 failures or 2060 hours, whichever occurs first. Failure truncation will result in reject-decision and time truncation will cause accept-decision.

Standard PRST Tests

The MIL-HDBK-781 provides eight standard sequential plans which could be used readily instead of having to design tests as detailed above. The eight include six basic plans designated as I-D through VI-D (the suffix D comes from the fact that the MIL-HDBK-781 is an appendix to the MIL-STD-781D, and the plans are indexed with D to distinguish them from those given in earlier revisions of the standard) and two high-risk plans designated as VII-D and VIII-D. The basic plans are the recommended plans except that the short-run, high-risk plans can be used when overriding time and cost considerations warrant their use. The details of the standard plans are given in Table 7.2.

The accept-reject criteria for these eight plans have been worked out and are presented graphically in the military standard. These graphs are reproduced as Figures A.2(a) to (h) in the Appendix. The table below each graph is the numerical representation of the graph.

The x-axis on the graph represents total test time as multiple of lower test MTBF(Θ_1) and the y-axis represents chargeable failure. (A chargeable failure is a failure for which the manufacturer or the supplier is responsible.) The graph also shows an expected decision point for equipment that has true MTBF equal to the upper test MTBF (Θ_0).

The operating characteristic curves and expected test time for the eight plans are reproduced in the Appendix as Figures A.3(a) to (h).

Test plan	Nominal Risks (%) α & β	True Risks (%) α	β	Discrim. Ratio	Trunc Time Multiple of Θ_1
I-D	10	11.5	12.5	1.5	49.5
II-D	20	22.7	23.2	1.5	21.9
III-D	10	12.8	12.8	2.0	20.6
IV-D	20	22.3	22.5	2.0	9.74
V-D	10	11.1	10.9	3.0	10.35
VI-D	20	18.2	19.2	3.0	4.5
High-risk plans					
VII-D	30	31.2	32.8	1.5	6.8
VIII-D	30	29.3	29.9	2.0	4.5

Table 7.2 · *Standard PRST Test Plans and Their Characteristics*

Using the Standard PRST Plans

Although a sequential plan can be designed for any chosen set of risks and the discrimination ratio, availability of the standard test plans makes the selection of a sequential plan simple. Table 7.2 summarizes the characteristics of the eight standard plans given in the military standard. It is clear that when the values for α and β are small, or when the discrimination ratio is small, the test time will be long. Therefore, the selection of a sequential test plan has to be made based on the tradeoff between cost of testing and the risks acceptable in rejecting good equipment and that of accepting poor equipment.

The starting point in selecting a test plan is to decide the acceptable MTBF (Θ_0). This will be determined by reliability goals for the equipment. If it is a component of a larger system, it will be determined by the process of apportioning the reliability goals for the end system. Next, the value of d must be chosen. The smaller the value of d, the larger the value of Θ_1, hence less chance of accepting products with poor reliability. Then, however, the smaller the d, the larger will be the test duration, hence the cost. A tradeoff is necessary.

One approach would be to start with a plan which has $d = 2$ and nominal value of 20% for α and β risks. Among the available standard plans this represents the middle-of-the-road selection. The plan can be modified for future tests based on experience with the results from this plan.

Confidence Limits for MTBF

It is possible to obtain a confidence limit for MTBF of the equipment under test as a result of the PRST tests. The elaborate tables that are available in the handbook to obtain confidence intervals from PRST tests are not included in this text.

Program Manager's Assessment

The standard PRST plans have a provision by which the program manager can stop a test and accept the product when it is clear the test is heading for an accept decision. The plans, however, do not allow such early rejection when the test is heading toward a rejection. The reader is referred to the military handbook for details of such a provision.

Fixed Duration Tests

Fixed duration tests have the obvious advantage in test planning. These tests can be used for both qualification and production acceptance testing.

The fixed duration test consists of running the equipment for a total test time of T hours (or other life units such as cycles, miles, etc.) and accepting if the number of failures is a or less or rejecting if the number of failures is r or greater. The number of units to be tested in order to accumulate T hours must be decided from practical considerations. Again, whether it should be a replacement test when a failed part is replaced in the fixture or socket by a new one of the same type and the test continued or nonreplacement test, should be decided based on the equipment being tested, special fixtures needed, and similar practical considerations. On the assumption of exponential life, the two types of tests do not make a difference on the statistical criteria. The test is designed by selecting T, a, and r for given values of α, β, Θ_0, Θ_1.

Assuming exponential distribution for failure times (or Poisson distribution for number of failures in a given time T), a and r must be chosen to satisfy the following two equations:

$$\beta = \sum_{k=0}^{a} \frac{\left(\frac{T}{\Theta_1}\right)^k e^{-T/\Theta_1}}{k!} \qquad 1 - \alpha = \sum_{k=0}^{r-1} \frac{\left(\frac{T}{\Theta_0}\right)^k e^{-T/\Theta_0}}{k!}$$

The first equation makes the probability of acceptance β when the true MTBF of the equipment equals Θ_1, and the second equation makes this probability $(1-\alpha)$ when the true MTBF is Θ_0. In addition, the rejection number r is made equal to $a + 1$ in order that a decision is reached within the time T, i.e., if the equipment is not accepted in a test, it is rejected.

The test design is made by choosing a and r by trial and error, starting from $a = 0$, and continuing the trial until the equations are satisfied simultaneously. The following example illustrates the procedure.

Example

Design a fixed duration test to satisfy the following requirements: $\alpha = 0.2$; $\beta = 0.2$; $\Theta_0 = 1000$; $\Theta_1 = 500$

Solution

First select $a = 0 \rightarrow r = 1$

$$\beta = \sum_{k=0}^{0} \frac{(T/500)^0 e^{-T/500}}{0\,!} = 0.2$$

$$(1 - \alpha) = \sum_{k=0}^{0} \frac{(T/1000)^0 e^{-T/1000}}{0\,!} = 0.8$$

Recognizing the fact that the summation terms of both the equations are cumulative Poisson probabilities, cumulative Poisson tables (Table A.4) can be used to solve the equations. From Poisson tables, $T/500 = 1.6$ or $T = 800$ will satisfy the accept equation. Substituting this value of T in the second equation (reject equation) we obtain, again using Poisson table, for $\lambda = 800/1000$, $(1 - \alpha) = 0.45$, i.e., $\alpha = 0.55$. This value of α is too large. We try again with $a = 1$ and $r = 2$ to obtain $\alpha = 0.44$ which is still large. After a few more similar trials we obtain $\alpha = 0.2$ for $a = 5$, $r = 6$ and $T = 3900$.

The test plan would be: Test the equipment for total operating time of 3900 hours, accept if five or less failures are observed, and reject if six or more failures are observed.

The MIL-HDBK-781 gives several standard test plans identified as IX-D to XVII-D and XIX-D to XXI-D. The test plans along with their decision risks are reproduced as Table 7.3. The plans XIX-D to XXI-D are labeled as high-risk plans because of the larger α and β values and, as can be seen from the table, they entail shorter testing time. The OC curves for these test plans are shown in Appendix Figures A.4(a) to (l).

The handbook also has several "alternative fixed duration test plans" for several combinations of producer's risk and consumer's risk. With the availability of these, plans could be chosen to accommodate any time constraints while meeting risk requirements. Since the standard plans are considered to provide an adequate number of alternatives to choose from, these alternative plans are not included here. The readers are referred to the handbook if additional choices are necessary.

MTBF Estimation from Fixed Duration Tests

When a contract requires that a specified MTBF be met, the fixed duration test can be used. It must be pointed out that the lower test MTBF(Θ_1) and upper test MTBF(Θ_0) used for designing the test cannot

Test Plan	True Decision Risks (Percentage) α	β	Discrim Ratio (d)θ_0/θ_1	Test Duration (Multiples of θ_1)	Number of Failures: Reject (Equal or More)	Accept (Equal or Less)
IX-D	12.0	9.9	1.5	45.0	37	36
X-D	10.9	21.4	1.5	29.9	26	25
XI-D	19.7	19.6	1.5	21.5	18	17
XII-D	9.6	10.6	2.0	18.8	14	13
XIII-D	9.8	20.9	2.0	12.4	10	9
XIV-D	19.9	21.0	2.0	7.8	6	5
XV-D	9.4	9.9	3.0	9.3	6	5
XVI-D	10.9	21.3	3.0	5.4	4	3
XVII-D	17.5	19.7	3.0	4.3	3	2
High-risk fixed-duration test plans						
XIX-D	29.8	30.1	1.5	8.1	7	6
XX-D	28.3	28.5	2.0	3.7	3	2
XXI-D	30.7	33.3	3.0	1.1	1	0

Table 7.3 · *Summary of Standard Fixed Duration Tests*

be used to denote the inherent reliability of the equipment in question. In order to find the MTBF that expresses the capability of the equipment, the *demonstrated* MTBF has to be derived from test results. The demonstrated MTBF is expressed as a confidence interval. Since the formulas to be used for computing the confidence interval are different depending on whether the test resulted in an accept or a reject decision, the two cases are discussed separately. However, in both cases a point estimate of demonstrated MTBF ($\hat{\theta}$) is made by dividing the total test hours by the number of failures in that period.

Although formulas can be used to compute confidence interval for any level of confidence, a confidence level equal to $(1 - 2\beta)$ (where β is the value of consumer's risk used in designing the test) is recommended in computing the interval estimate.

Case 1: MTBF estimation at reject decision

Since the rejection decision occurs at the time of a failure, this is a failure-truncated test.

The confidence interval is obtained by multiplying the point estimate of MTBF ($\hat{\theta}$) by the following multipliers:

Multiplier for lower limit: $\dfrac{2r}{\chi^2_{(1-c)/2;\ 2r}}$

Multiplier for upper limit: $\dfrac{2r}{\chi^2_{(1+c)/2;\ 2r}}$

where r = number of failures, c = confidence level (in decimal fraction), and $\chi^2_{\alpha;\ \nu}$ = the value of the chi-square variable with ν degrees of freedom that cuts off α probability at the upper tail.

Case 2: MTBF estimation at accept decision

Since the accept decision occurs at the end of the test period, this corresponds to a time truncated test. The multipliers are given by the following formulas.

Multiplier for lower limit: $\dfrac{2r}{\chi^2_{(1-c)/2;\ 2r+2}}$

Multiplier for upper limit: $\dfrac{2r}{\chi^2_{(1+c)/2;\ 2r}}$

(Note the degrees of freedom for the χ^2 variable are different for the two multipliers.)

In order to facilitate the computation of the confidence intervals, the MIL-HDBK-781 presents tables and graphs that give the value of the above multipliers readily. These are reproduced in the Appendix as Table A.5 and Figure A.5 for the reject decision and Table A.6 and Figure A.6 for accept decision.

Example

(Reject Decision)

In a reliability test the seventh failure occurred at 820 hours of total test time. Find out an 80% confidence interval for demonstrated MTBF.

Solution

$$\hat{\Theta} = \frac{820}{7} = 117.14 \qquad c = 0.8$$

$$\frac{1+c}{2} = \frac{1+0.8}{2} = 0.9 \qquad \frac{1-c}{2} = \frac{1-0.8}{2} = 0.1$$

$$\chi^2_{0.9;\ 14} = 7.79 \qquad \chi^2_{0.1;\ 14} = 21.06$$

$$80\%\ \text{CI for MTBF} = \left[117.4 \times \frac{14}{21.06},\ 117.4 \times \frac{14}{7.79}\right]$$

$$= [78.04,\ 210.99]$$

$$= [78,\ 211]$$ (round MTBF estimates to the nearest integer)

Using the Multipliers from the Table

Enter Table A.5 with the number of failures and the percent confidence level required to obtain:

Multiplier for lower limit = 0.665

Multiplier for upper limit = 1.797.

These multipliers provide the same confidence interval as the formulas.

It should be remembered that the lower limit of the 80% CI gives the 90% (one-sided) lower bound for the MTBF and the upper limit gives the 90% (one-sided) upper bound for the MTBF.

The interpretation of these results will be: There is an 80% chance that the true MTBF lies between 78 and 211 hours. There is a 90% chance that the true MTBF is not lower than 78 hours. There also is 90% chance that the MTBF is not higher than 211 hours.

Example (Accept Decision)

A test resulted in an accept decision after 920 hours with seven failures during that period. Find an 80% confidence interval for the demonstrated MTBF.

Solution

$$\hat{\theta} = \frac{920}{7} = 131.4 \text{ hrs} \qquad c = 0.8$$

$$\frac{1 + c}{2} = 0.9 \qquad \frac{1 - c}{2} = 0.1$$

$$\chi^2_{0.9;\,14} = 7.79 \qquad \chi^2_{0.1;\,16} = 23.54$$

$$\text{Lower limit multiplier} = \frac{14}{23.54} = 0.595$$

$$\text{Upper limit multiplier} = \frac{14}{7.79} = 1.797$$

$$\begin{aligned} 80\% \text{ CI for MTBF} &= [131.4 \times 0.595,\ 131.4 \times 1.797] \\ &= [78,\ 236] \end{aligned}$$

The tabled values for the multipliers can be seen to agree with the values calculated from the formulas.

Provision for Early Acceptance

The fixed duration plans given in the military handbook include a provision for early acceptance (not waiting until the fixed duration time) especially when the equipment has high reliability. Appendix Table A.7, reproduced from the handbook, gives for the standard fixed duration plans the accept times T_j such that the test can be terminated with an accept decision if no more than j failures have occurred up to T_j time units. The T_j's are given in multiples of the lower test MTBF(Θ_1).

For example, the standard test plan IX-D calls for running the test up to $45\Theta_1$ time units and accepting only if no more than 36 failures occur in that time period. The provision of early acceptance means that this test can be terminated with an accept decision if no failure occurs within $4.2\Theta_1(T_0 = 4.2)$ time units, or if no more than one failure occurs within $6.1\Theta_1$ ($T_1 = 6.1$) time units, etc.

Such early acceptance decision changes the OC of the standard plans, but only moderately. Such early decisions, however, contribute to saving in test costs.

Minimum MTBF Assurance Tests

The MTBF assurance test is used when it is needed to provide assurance that a minimum MTBF level is achieved in the equipment. This test can be used on production equipment which has passed previous qualification tests. An environmental stress screening must precede this test in order to eliminate the early failures. These tests are designed to provide a specified (high) probability of acceptance when the equipment has the MTBF equal to or larger than a specified minimum MTBF.

These tests require that the equipment operate without failure for a certain failure-free interval of t hours within a test window of W hours. In other words, if $W = 200$ and $t = 100$, the equipment under test should operate for at least 100 hours without a failure, within a 200-hour test period, in order to be accepted. The test parameters W and t are chosen such that the probability that an equipment with MTBF equal to or larger than the minimum MTBF is a given value P_s. It can be shown (MIL-HBBK-781D contains proof) that the parameters W and t can be selected using the following relationship:

$$P_s = \frac{(M - 1)^t \, (M + W - t)}{M^{t + 1}}$$

where M = Minimum MTBF, hours
W = test window, hours ($t \leq W \leq 2t$)
t = failure free interval

Statistical significance considerations recommend the use of $W = 2t$; hence the above relationship reduces to:

$$P_s = \frac{(M - 1)^t \, (M + t)}{M^{t + 1}}$$

Therefore, for any value of P_s, the probability of acceptance desired for acceptable equipment, the value of t and hence the value of W can be obtained. This equation can be solved for t, given P_s and M using numerical methods. Appendix Figures A.7 and A.8 give values for t for several values of P_s and minimum MTBFs.

Example

Suppose a test is needed to assure a minimum MTBF of 600 hours and probability of acceptance for equipment having MTBF larger than this minimum should be 0.98. From Appendix Figure A.7, value of $t = 130$. Then $W = 260$.

The test procedure would be to run the equipment under normal operating conditions until a stretch of 130 hours of failure-free operation or 260 hours, whichever occurs first. If a stretch of 130 hours of trouble free operation is obtained before 260 hours, the equipment is accepted as satisfying the minimum MTBF of 600 hours; otherwise rejected.

The MTBF assurance test is most suitable to test production equipment, especially when each piece of equipment has to pass a reliability test.

It also is possible to use the data from these tests to estimate the MTBF achieved based on the accumulated data on all tested equipment or on a group of latest equipment. The values of t and W can then be adjusted depending on whether the estimated MTBF is worse than or better than the minimum MTBF. If the most recent data indicate a significant decrease in MTBF compared to the required minimum, the test window as well as failure-free time can be increased to reduce the probability of accepting unacceptable equipment. If this results in improvement in MTBF the test parameters can be adjusted to near original levels.

All-Equipment Production Reliability Acceptance Tests

This type of test is used when every unit of production has to be tested for reliability. The MTBF assurance tests described in the previous section would be suitable for this purpose. However, a sequential test plan which is a modification of the standard PRST plan III-D is offered as the basic plan for all equipment reliability testing. This plan designated as XVIII-D is shown in Appendix Figure A.9.

Each unit of equipment must be run for a minimum of 20 hours and a maximum of 50 hours. Where length of test is counted by number of cycles, the test should be ended with the end of the cycle after the predetermined time is completed.

When the units are tested in their normal environmental conditions, the cumulative failures and the cumulative equipment operating times are recorded. The cumulative number of failures and cumulative test times are plotted on Appendix Figure A.9. A decision is made according to where the plots fall in the graph.

If the plots reach the reject region, the test will be stopped and the production lot is declared not acceptable; the test should be terminated and corrective action undertaken. If the specified test time is completed without reaching the reject line, all of the equipment which the lot under the test comprises will be considered acceptable.

It also is possible to develop all equipment test plans out of any standard PRST plans shown graphically in Figures A.2(a) to (h). However, a slight modification to these plans is required arising out of the fact that these plans had been originally modified to take into account the effects of truncation. The all-equipment plans are the unmodified sequential plans. Such all-equipment plans corresponding to each of the standard PRST plans are available in the handbook.

Environmental Stress Screening (ESS) Tests

The ESS tests are conducted to ensure that an equipment that has been previously qualified by reliability testing has not suffered loss of reliability due to production deficiencies causing early failures. The detection and elimination of early failures is accomplished by subjecting the equipment to environmental conditions more severe than those encountered in normal operating conditions.

The type and severity of environmental stress used for electronic equipment (e.g., thermal stress, vibration stress, moisture, electrical stress, and duty cycle) are described in detail in MIL-HDBK-781. Some of these same types of environmental stress may be applicable to non-electronic equipment as well, but the particular types of environment stresses suitable for a given piece of equipment cannot be specified herein. They must be selected by the test engineer. For example, increased pressure and temperature may be suitable while testing hydraulic hoses, higher-impact loads may be used while testing door locks; and overloads may be appropriate for testing electric motors.

Once the particular type of environmental stress has been selected, the length of time such stress will be imposed must be determined. There are three methods described in the handbook that may be suitable for making this time determination.

1. The computed ESS time interval method

   This analytical method provides a technique for estimating the required ESS time to ensure that, with a high probability, the defective parts in a system (or equipment) will have been discovered. This method requires data on the total number of parts in the system, the probability that a part is defective, and the failure rate of each defective part. This technique is not described in any detail in this text and the reader is referred to MIL-HDBK-781 for further information.

2. Graphical method

In this method, data are collected from the ESS test results and a plot of observed failure rate (number of failures/operating hours) is made to approximate the actual failure rate curve as shown in Figure 7.4. Typically these curves will flatten out if the defective parts are removed or repaired by ESS. The ESS duration is obtained by observing when the curve becomes flat. For example in Figure 7.4 an ESS duration of approximately 70 hours would be reasonable. Therefore, each unit of the product in question will be run for 70 hours under the chosen environmental conditions to ensure elimination of early failures. For equipment that does not have any prior historical data the ESS duration can be chosen based on data for similar equipment and then revised based on accumulated data from the tests.

3. Standard ESS

The MIL-STD-2164 "Environmental Stress Screening Process for Electronic Equipment" contains a standard ESS plan for electronic equipment. It recommends the appropriate levels of thermal and vibration stresses to be applied and the duration for the *pre-defects free* test and the *defects free* test. In other words this test requires a defect-free test period of a certain number of hours under the stress conditions after the equipment has been subjected to the same stress levels for a period when defects could occur and be rectified. The standard also allows for a sampling plan under certain conditions, where not all the production units are subjected to ESS test but only a sample chosen according to a sampling plan. For details of the procedure the reader is referred to MIL-STD-2164.

Out of the three methods available for ESS, the graphical method is the simplest and can be applied to any type of equipment.

Summary on Test Plans

Several test plans from the MIL-HDBK-781 discussed in this chapter are available to meet different needs. The following is a summary of these plans, their objectives, and the purpose for which they can be used.

The *Duane plot* is used for reliability development and growth testing (RDGT) when the test-analysis-and-fix sequence will result in improvement to the product design and hence will result in growth in reliability. The Duane plot will model the growth and enable evaluation of reliability attained (MTBF) at the end of the growth testing. Generally, a numerical goal for reliability is set and RDGT is done until the goal is reached or exceeded. RDGT is best performed on the first one or

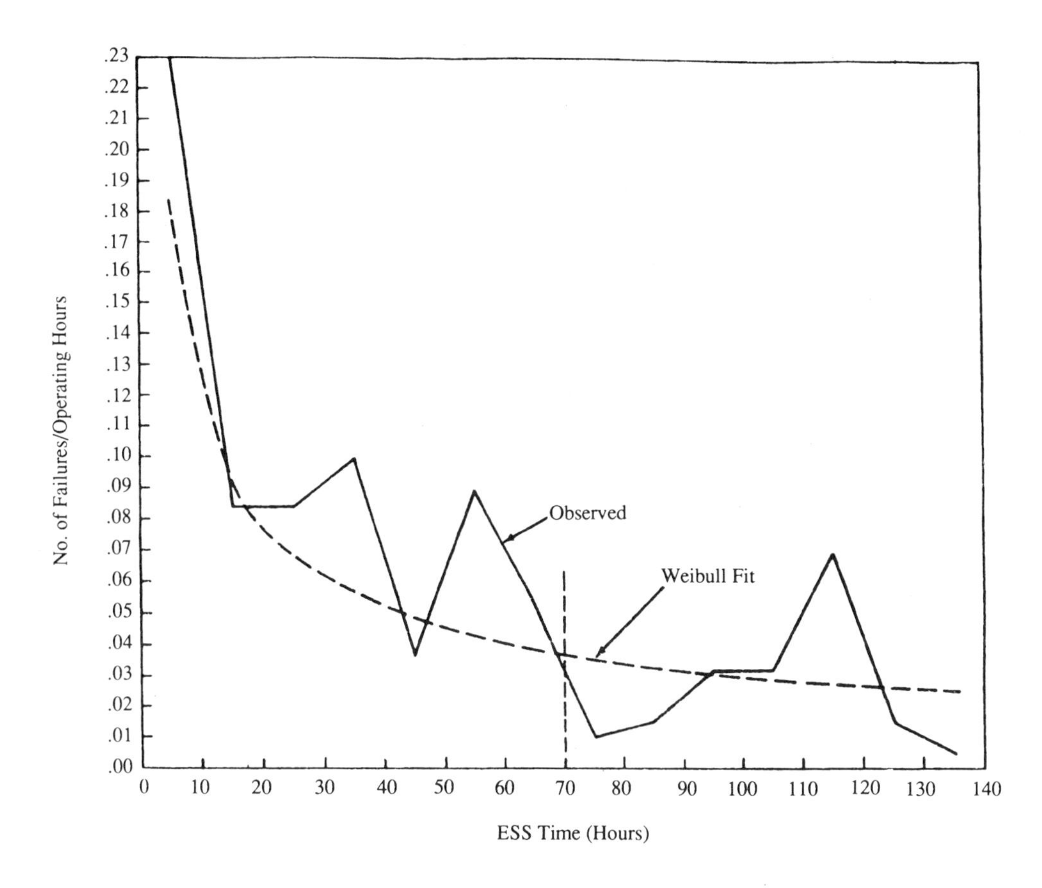

FIGURE 7.4 · *Example of graphical data plot for ESS*

two prototypes of a product so that the information from the test can be put to best use in making appropriate design changes.

The ESS (environmental stress screening) is done in order to weed out early failures and incorporate corrective action to avoid recurrence of such failures. These failures are evoked by subjecting the product to higher stress levels than they would normally be subjected to in actual service. ESS is recommended on 100% of production in the early part of the production cycle, to be reduced when the failure rate in the population becomes constant. ESS must be completed on items before they are subject to RDGT, qualification, or acceptance testing. The graphical method is recommended to determine the time period for ESS tests.

RDGT and ESS tests are referred to as *reliability engineering tests* since they are useful in improving the reliability of a product. In comparison with the qualification and acceptance tests, these have the most return on investment and must be the choice when reliability improvement is the goal.

The probability ratio sequential test (PRST) plans are intended only to determine compliance of a product to reliability specifications defined in terms of decision risks. These are used for qualification and production acceptance tests. Because of the uncertainty associated with test duration, however, these are not the preferred tests when test time is important from a budgeting point of view, and results are necessary before certain time limits. The fixed duration tests then become the preferred choice. These require selection of a fixed duration (T) and an acceptable number of failures (a). The test consists of running the equipment for T hours and if no more than a failures occur, the equipment is accepted; otherwise it is rejected. These tests can be used for both qualification and production acceptance.

The MTBF assurance tests are suitable primarily for production acceptance. These require selection of a test time window (W) and a failure-free interval (t). The equipment will be accepted if it runs failure free for t hours within a test period of W hours. The fact that the test has a maximum time limit and that MTBF estimate can be obtained from test results are the advantages. This also can be used for all equipment testing. The all-equipment test plan XVIII-D, which is a modification of the standard PRST plan, is used when a PRST plan is needed to test all equipment.

CHAPTER 8

A System for Reliability

Several statistical methods necessary for defining, measuring, and making predictions about reliability were discussed in earlier chapters. Besides these statistical methods, there are a few other organized approaches that have been used in accomplishing reliability goals. These are covered in this chapter. Once again one of the military standards MIL-STD-785B is used as a basic resource. The military standard describes a system and its various components that must be in place in order to achieve reliability in products. Some of the important components of the system are described below.

Reliability Goals

Reliability goal setting is the first item on the agenda of a reliability program. This goal is chosen for the total system at the highest level. For example, a diesel engine manufacturer will choose the reliability for the engine, and a truck manufacturer will choose the reliability goal for the total truck.

The goal selection should be based on prior history, a competitor's record or demands of the customer, and reasonable judgment of what can be accomplished. The total cost of developing, manufacturing, operating, and maintaining the equipment, or the life-cycle cost, would also influence the selection of the reliability goal. The goal should be stated along with specification for performance requirement of the product and the expected environmental conditions.

Apportionment

The reliability goal chosen for the total system then should be apportioned to the next level of subsystems. This apportionment is an iterative process completed in order to obtain subsystem goals that would reasonably match subsystem capabilities. This is accomplished by first preparing the reliability model or block diagram (RBD) that would reflect the functional relationship of the subsystems to the system function. Of course, at any stage in the development of the product, the model would only reflect the current level of details achieved in the design process. The model can be updated as the design process progresses. The statistical methods discussed in Chapter 6 are used to obtain system reliability from subsystem reliabilities. The subsystem reliabilities at this stage are the estimated values, or those provided by subsystem suppliers or published values for off-the-shelf items.

A few iterations are necessary in order to allocate the system goal to the subsystems in such a way that larger reliabilities are required of more capable subsystems in order to accommodate the less capable subsystems. More often than not it will become apparent during this process of allocation, that at least for some of the subsystems the capability falls short of the requirements. These subsystems will be identified as *reliability critical* items. These are the items that should be subjected to further analysis, testing, and fixing to improve their reliability to meet the goals required of them.

Discussion among the reliability group and design engineers should follow in order to find ways of improving the reliability of critical items. Such discussions have the salient benefit of creating an awareness among the group for the problem's existence and extent. At the end of such discussions a recommendation should come for the plan to improve the reliability of the critical items, along with a rough estimate of the extent of work needed, and the cost of accomplishing the goal.

At some point in the process of allocating goals to the subsystems, the apportionment must be frozen and the reliability requirements on the subsystems levied on the subsystems' designers or suppliers. These should form the basis for further allocation of reliability requirement for assemblies and parts.

The Reliability Organization

For any reliability program to be successful there should be a reliability organization within a company, staffed by qualified reliability engineers, with responsibility to plan, initiate, monitor, and coordinate reliability activities. There seems to be an agreement among authorities that the engineers who are familiar with the design and manufacturing phases of a product when trained in reliability methods, will be best suited to perform reliability functions. Help from statisticians with advanced knowledge can be sought when the complexity of the reliability problems warrant such help. The reliability group should be adequately staffed and budgeted commensurate with the amount of work involved.

Where should the reliability unit be located in an organization? Traditionally they are located either within the quality group or the design engineering group. The quality organization is made up of people with skills and expertise similar to those needed for reliability work. Locating within the engineering group has the advantage that reliability work has the most to do with engineering and the engineering can benefit the most from reliability work. There also is an argument that reliability should be the responsibility of a group independent of quality and engineering since they will have the ability to critically evaluate both of these activities which are the major contributors to reliability.

Where the reliability unit is located in an organization depends on the size of the organization, extent of reliability work needed, current availability of expertise, and the culture of the organization with respect to the ability of the groups to work with one another. Irrespective of where the unit is located, it is important that there is a reliability unit that must include a manager entrusted with responsibility for coordination of work relative to reliability, and persons with capabilities in engineering and statistical methods who can work with people engaged in overlapping activities in reliability.

Not all production organizations need extensive reliability departments. A company making wood screws or producing iron castings will not need a reliability department because there are no complex life-relationships between parts and assemblies; there are proven designs and the cost of failure is not enormous. On the other hand, a factory producing an automobile fuel pump, brake system, or transmission will have to be concerned about reliability and must have the expertise suitably assembled to perform functions required to achieve reliability goals.

Design Reviews

Design reviews are the milestones in a reliability program mcant to monitor the progress toward reliability goals, made by a group of people having responsibility for reliability. Typically the formal design review takes place at two stages. The first stage is to verify and approve the adequacy of the broad selection of design criteria at preliminary design stages before detailed designs are undertaken. The second stage is to verify and approve the detailed design before commitment to production.

In the first stage the choices are between major types of components (e.g., between electronic and conventional ignitions, or between disk and drum brakes, etc.) The reliability analysis at this stage will be only a paper analysis based on the known reliabilities of components from past history. The second stage will have more detailed design, ready for production, and may include results of testing with prototypes.

The review team is comprised of people from engineering, production, quality assurance, reliability, sales, and possibly customer representatives. The review is not to be interpreted as second-guessing the design engineers; it is a method of guaranteeing that adequate attention has been paid to reliability aspects while satisfying technical and performance requirements. Although the design engineers have the primary function in the designs, an independent review by the group assures that producibility, cost, weight, safety, packaging, and handling, etc., also have been given proper consideration.

Besides the formal review events, more continuous informal reviews by the reliability engineer and the design engineer, working together, help to resolve many reliability issues such as choice of parts, configuration, and tests to be performed. They also contribute to reaching prompt decisions during the formal review sessions.

Review groups who have experience with a particular type of product generate checklists for design reviews for use at formal review sessions. No one checklist can be used for all product types. The checklists can be developed by a given group as they gain experience with a product or a family of products.

Design reviews are important landmarks in the product development cycle and are meant to ensure that reliability considerations are given adequate attention at the time when it is most appropriate to make changes, if deficiencies are discovered.

Failure Reporting, Analysis, and Corrective Action System (FRACAS)

The FRACAS is the way to ensure that failures encountered during developmental testing are properly reported, causes identified, and corrective actions implemented so that these failure modes do not recur either in further testing or during later service with the customer. The closed loop system of reporting, analysis, and corrective action usually is supervised by a failure review board (FRB) consisting typically of the design, reliability, production, and quality engineers.

Essential to the effectiveness of the FRACAS is a failure reporting form which would include the following information:

- Symptoms of the failure and its effect
- Conditions surrounding the failure (load, temperature, humidity, etc.)
- Failure hardware identification (for traceability)
- Operating time (cycles, mileage, etc.) at failure
- Date/time of failure

- Identification of failure mode
- Analysis of all part failures
- Repair action taken to return equipment to operation
- Corrective action necessary to prevent recurrence of such failure

A failure report should be initiated at each failure and the failed equipment or part should be properly tagged with the serial number of the failure report pertaining to the item written on it.

The FRACAS also should provide for periodic review of the closed loop system to make sure that stages of reporting, analysis, and corrective action for each of the failures are adequately followed.

Reliability Predictions

Just as apportionment or allocation is a topdown approach to determine reliability requirements of subsystems, assemblies, subassemblies, and parts, prediction is the bottom-up method for assessing the reliability of the entire system from the reliabilities of parts and assemblies, as the design progresses and data become available on part reliability. Predictions falling short of requirements at any level should indicate need for attention. Simplification of design, higher-quality parts, improved environment, redundancy, etc., may have to be tried in order to improve the predicted reliability. On the other hand, a serious shortfall might point to the need for another iteration of the reliability allocation to better match the requirement and prediction.

Predictions should be made as early as possible and updated as the design details are being decided. Although the early predictions are unrefined, they nevertheless provide information to determine the reliability feasibility of the system at that stage. As the design progresses and paper designs are converted into parts and subassemblies, data from reliability tests can be incorporated in the predictions and predictions become more reliable.

Predictions should not be taken as the reliability achieved in the equipment, the reliability achieved in the equipment can be obtained only from test results.

Failure Mode, Effects, and Criticality Analysis (FMECA)

FMECA is a powerful tool in the hands of a reliability engineer. It is used to discover potential design weaknesses and rectify them at the design stage. The method consists of identifying the likely modes of failure, the possible effects of each failure and the criticality of each effect on reliability of the product. A reliability criticality number is assigned to each failure mode based on probability of occurrence, severity of the failure effect, and the chance of being detected either during production or during inspection and testing before reaching the customer. These criticality numbers are then used to assign priority for corrective action such as adding redundance, providing environmental protections, devising preventive maintenance schedule, or planning for additional growth testing. Figure 8.1 shows an example of an FMECA analysis.

FMECA should be completed early in the concept stage, which offers the most potential for design changes, and should be repeated as design details become more defined. At the concept stage the FMECA may disclose some of the single point failures (failure modes that cause failure of the system or the entire assembly) that can be eliminated by simple rearrangement of the schematics. At the advanced design stage the FMECA can be performed up to the parts level to study the effect of part failures on the total system.

Because of input required from various disciplines for successful use of FMECA analysis, it is important that it be done by a team which brings together knowledge and skills from various disciplines such as design, production, quality control, purchasing, and reliability engineering.

Parts Selection/Application Program

The parts selection/application program has the objective of maximizing the selection and use of *standard parts.* A parts control program increases the probability of achieving and maintaining equipment reliability because of the use of parts with known quality and reliability, minimizes parts proliferation and thus minimizes parts inventory costs. Experience has shown that the added investment needed for the parts control program is sufficiently paid back by improved reliability and reduced life-cycle cost of the equipment.

A parts control program will consist of the following:

1. A documented procedure that formalizes the program
2. Parts standardization (with a standard list)
3. Parts application guideline (load, environment, etc., conditions including derating* procedures)
4. Parts testing, screening, or approval procedure

When a nonstandard part is proposed for selection it should be approved only after all options to include standard parts have been investigated.

*Derating is the method of selecting a part for conditions less severe than its rated conditions so as to obtain improved reliability.

FAILURE MODE, EFFECTS, AND CRITICALITY ANALYSIS

System:__________ Suppliers:__________ Prepared:__________ Date:_____

Model Year:__________ Scheduled Date:__________ Approved:__________ Date:_____

Part/Process Name Part Number	Part/Process Function	Failure Mode	Effect(s) of Failure	Cause(s) of Failure	Occur.[1]	Sever.[2]	Detect.[3]	Risk Prior No.[4]	Recommended Corrective Action & Status
Capacitor	Timing capacitor to shut down speed if brake switch fails	1. Shorted	• Uncontrolled accel. • System will not turn off • Brake disable switch inoper.	• Pin holes in dielectric • Inadequate connection	1	10	10	100	Relocate capacitor in circuit to produce fail-safe condition.
Master cylinder	Transmit brake application to wheel cylinders	1. Leakage	• Loss of fluid • Loss of brake effectiveness • Loss of brake	• Bad seal	2	10	6	120	Reduce variation in seal thickness Inform supplier. Increase surveillance

1. Frequency of occurrence: 1—infrequent, 10—very frequent
2. Severity: 1—minor inconvenience, 10—serious damage to equipment, or operator
3. Detectability: 1—very likely the defect will be detected before reaching customer, 10— very unlikely
4. Product of the scores for above 3—larger the more critical

FIGURE 8.1. · *Example of FMECA analysis*

Reliability Testing

Reliability testing was covered in detail in Chapter 7. It is discussed again to emphasize its part as a component of the reliability system. As discussed in Chapter 7, testing is used to evaluate reliability of products at various stages: at the development stage to disclose deficiencies in design and to monitor growth in reliability, for qualification prior to committing a design to production, after production to demonstrate meeting of reliability specifications, and for eliminating the possibility of weak parts affecting the reliability of the equipment through early failures.

Several test procedures have been devised to meet the different objectives mentioned above. Selection of proper testing procedure according to the needs of the occasion would help in achieving maximum information with minimum testing. Although prediction exercises provide useful information on reliability for purposes of estimating to study feasibility or make apportionment, only testing can prove the inherent reliability achieved in a product. Hence testing serves as an important component of a reliability system.

Wherever possible, reliability testing should be combined and integrated with performance testing to economize on testing costs. As far as possible the test environment should be realistic in the sense it simulates the conditions in which the equipment ultimately will be working. Where possible the testing should be done by a group independent of production to avoid bias in the evaluations. It is essential that complete records be kept on the test plans, schedules, failures observed, corrective actions taken, data analysis, and final results.

Reliability System Summary

A systematic approach is necessary to initiate, monitor, and sustain activities to accomplish reliability goals. An organization staffed with people qualified to perform reliability work is necessary to set goals, keep schedules, and coordinate activities relative to reliability. Design reviews provide the necessary check at the appropriate stages of product development to ensure that reliability is given proper consideration along with performance, producibility, cost, etc. A failure reporting system is needed to ensure that design corrections are incorporated to avoid failures encountered in testing. A parts selection program ensures use of standard approved parts to avoid introduction of parts of unknown quality and reliability. The testing program is necessary to verify the results of reliability effort. There also are other activities such as failure-tolerant design, sub-contractor's supplies control and sneak circuit analysis that would be of interest in particular situations. We have included here only those components that most reliability systems would require.

It is clear that the efforts mentioned above as part of a reliability system are necessary in order to achieve reliability objectives in products. These procedures also guarantee that decisions relative to reliability are not left to the judgment of one or two individuals, but are subject to the collective effort of a group of professionals concerned with reliability.

References

1. Dhillon, B. S. *Reliability Engineering in Systems Design and Operation*, Princeton, N.J.: Van Nostrand Reinhold, 1983.

2. Duncan, A. J. *Quality Control and Industrial Statistics*, 4th ed. Homewood, IL: Richard D. Irwin, 1974.

3. Duane, J. J. "Learning Curve Approach to Reliability Modeling," *IEEE Trans Aerospace,* 2, pp 563–566, 1964.

4. Hines, W. W. and Montgomery, D. C. *Probability and Statistics in Engineering and Management Science,* 3rd Ed. New York, John Wiley, 1990.

5. Ireson, W. G. and Coombs C. F. (Eds.), *Handbook of Reliability Engineering and Management*, New York: McGraw-Hill, 1988.

6. Kapur, K. C. and Lamberson, L. R. *Reliability in Engineering Design*, New York: John Wiley, 1977.

7. Lewis, E. E. *Introduction to Reliability Engineering,* New York: John Wiley, 1987.

8. Lloyd, D. K. and Lipow, M., *Reliability: Management Methods and Mathematics*, 2nd ed., Milwaukee, WI: ASQC, 1984.

9. MIL-STD-781D. "Reliability Testing for Engineering Development, Qualification, and Production," 1986.

10. MIL-HDBK-781. "Reliability Test Methods, Plans, and Environments for Engineering Development, Qualification and Production," 1987.

11. MIL-STD-785B. "Reliability Program for Systems and Equipment Development and Production," 1980.

12. MIL-STD-2164. "Environmental Stress Screening Process for Electronic Equipment," 1985.

13. MIL-STD-189. "Reliability Growth Management," 1981.

14. MIL-STD-1629A. "Procedures for Performing a Failure Mode, Effects, and Criticality Analysis," 1980.

15. Nelson, W. *Applied Life Data Analysis*, New York: John Wiley, 1982.

16. ______. "Theory and Applications of Hazard Plotting for Censored Failure Data," *Technometrics*, 14, 1972.

17. O'Connor, P. D. T. *Practical Reliability Engineering*, 2nd ed., New York: John Wiley, 1985.

18. Smith, C. O., *Introduction to Reliability in Design*, New York: McGraw-Hill, 1976.

19. Tobias, P. A. and Trindade, D., *Applied Reliability*, New York: Van Nostrand Reinhold Co., 1986.

20. Walpole, R. E., and Myers, R., *Probability and Statistics for Engineers and Scientists*, 4th ed., New York: Macmillan, 1989.

Answers to Selected Exercises

Chapter 1 1.3) 3/6 1.4) 3/10 1.5) 19/36 1.6) 210/216 1.7) 0.2048 1.8) 1.71% 1.9 a) 256 b) 1/256 1.10) 144 1.11a) 60 b) 24 1.12) 6/21 1.13 a) 0.662 b) 0.0354 c) 0.9984

Chapter 2 2.1)

y	2	3	4	5	6	7	8	9	10	11	12
p(y)	1/36	2/36	3/36	4/36	5/36	6/36	5/36	4/36	3/36	2/36	1/36

2.2) p(x) = 1/2, x = 1,2 2.3) 16/27 2.4) 7, 5.83 2.5) 1.5, 0.25 2.6) 0.9163 2.7) 0.423 2.8) 0.8413, 0.3446, 0.5565 2.9) 3.316 2.10) 0.41, 9.59 2.11) 0.6591 2.12 a) 0.0 b) 10.7 ounce

Chapter 3 3.1) [179.15, 180.85] 3.2) [1.1403, 1.1597] 3.3) [6.2202, 6.2478], [0.00033, 0.00154], [0.0182, 0.0393] 3.4) H_0: $\mu = 90$, H_1: $\mu < 90$, Do not reject H_0. Yield is not less than 90. 3.5) H_0: $\mu = 125$, H_1: $\mu > 125$, Do not reject H_0; mean life of batteries is not greater than 125 months. 3.6) H_0: $\mu_1 - \mu_2 = 0$, H_1: $\mu_1 - \mu_2 \neq 0$, Reject H_0, the machines are not filling equal volumes.

Chapter 5 5.1a) 0.0059 b) 0.988 c) 1941 5.3) 0.86 5.4) 0.0124 failures/day, 80.6 days 5.5) 0.016 failures/day, 62.1 days 5.6) 10,667 hrs 5.7) 0.187 failures/day, 5.34 days 5.8) [0.0065, 0.0216] failures/day, [46, 154] days 5.9) [0.0087, 0.0251] failures/days, [40, 115] days 5.10) [6,461, 19,049] hrs 5.11) 95% lower bound for MTTF = 126 hrs. Pumps do not meet spec. 5.12) 100 hrs 5.13) Yes, Exponential, 11,000 hrs 5.14) $\beta = 0.47$ $\Theta = 27$ hrs 5.15a) 0.512 b) 0.136

c) 0.0000093, 0.000064, 0.00015 5.16) $\beta = 2.1$ $\Theta = 110$ cycles
5.17) 0.6

Chapter 6 6.1) 0.66 6.2) 0.784 6.3) 0.342 6.4) 0.7168

Appendix

Normal Tables: gives F(z) = P(Z </= z), where Z ~ N(0,1)

z	0.00	0.01	0.02	0.03	0.04	0.05	0.06	0.07	0.08	0.09
-3.4	0.00034	0.00033	0.00031	0.00030	0.00029	0.00028	0.00027	0.00026	0.00025	0.00024
-3.3	0.00048	0.00047	0.00045	0.00043	0.00042	0.00040	0.00039	0.00038	0.00036	0.00035
-3.2	0.00069	0.00066	0.00064	0.00062	0.00060	0.00058	0.00056	0.00054	0.00052	0.00050
-3.1	0.00097	0.00094	0.00090	0.00087	0.00085	0.00082	0.00079	0.00076	0.00074	0.00071
-3.0	0.00135	0.00131	0.00126	0.00122	0.00118	0.00114	0.00111	0.00107	0.00104	0.00100
-2.9	0.0019	0.0018	0.0018	0.0017	0.0016	0.0016	0.0015	0.0015	0.0014	0.0014
-2.8	0.0026	0.0025	0.0024	0.0023	0.0023	0.0022	0.0021	0.0021	0.0020	0.0019
-2.7	0.0035	0.0034	0.0033	0.0032	0.0031	0.0030	0.0029	0.0028	0.0027	0.0026
-2.6	0.0047	0.0045	0.0044	0.0043	0.0041	0.0040	0.0039	0.0038	0.0037	0.0036
-2.5	0.0062	0.0060	0.0059	0.0057	0.0055	0.0054	0.0052	0.0051	0.0049	0.0048
-2.4	0.0082	0.0080	0.0078	0.0075	0.0073	0.0071	0.0069	0.0068	0.0066	0.0064
-2.3	0.0107	0.0104	0.0102	0.0099	0.0096	0.0094	0.0091	0.0089	0.0087	0.0084
-2.2	0.0139	0.0136	0.0132	0.0129	0.0125	0.0122	0.0119	0.0116	0.0113	0.0110
-2.1	0.0179	0.0174	0.0170	0.0166	0.0162	0.0158	0.0154	0.0150	0.0146	0.0143
-2.0	0.0228	0.0222	0.0217	0.0212	0.0207	0.0202	0.0197	0.0192	0.0188	0.0183
-1.9	0.0287	0.0281	0.0274	0.0268	0.0262	0.0256	0.0250	0.0244	0.0239	0.0233
-1.8	0.0359	0.0352	0.0344	0.0336	0.0329	0.0322	0.0314	0.0307	0.0301	0.0294
-1.7	0.0446	0.0436	0.0427	0.0418	0.0409	0.0401	0.0392	0.0384	0.0375	0.0367
-1.6	0.0548	0.0537	0.0526	0.0516	0.0505	0.0495	0.0485	0.0475	0.0465	0.0455
-1.5	0.0668	0.0655	0.0643	0.0630	0.0618	0.0606	0.0594	0.0582	0.0571	0.0559
-1.4	0.0808	0.0793	0.0778	0.0764	0.0749	0.0735	0.0722	0.0708	0.0694	0.0681
-1.3	0.0968	0.0951	0.0934	0.0918	0.0901	0.0885	0.0869	0.0853	0.0838	0.0823
-1.2	0.1151	0.1131	0.1112	0.1093	0.1075	0.1056	0.1038	0.1020	0.1003	0.0985
-1.1	0.1357	0.1335	0.1314	0.1292	0.1271	0.1251	0.1230	0.1210	0.1190	0.1170
-1.0	0.1587	0.1562	0.1539	0.1515	0.1492	0.1469	0.1446	0.1423	0.1401	0.1379
-0.9	0.1841	0.1814	0.1788	0.1762	0.1736	0.1711	0.1685	0.1660	0.1635	0.1611
-0.8	0.2119	0.2090	0.2061	0.2033	0.2005	0.1977	0.1949	0.1922	0.1894	0.1867
-0.7	0.2420	0.2389	0.2358	0.2327	0.2296	0.2266	0.2236	0.2206	0.2177	0.2148
-0.6	0.2743	0.2709	0.2676	0.2643	0.2611	0.2578	0.2546	0.2514	0.2483	0.2451
-0.5	0.3085	0.3050	0.3015	0.2981	0.2946	0.2912	0.2877	0.2843	0.2810	0.2776
-0.4	0.3446	0.3409	0.3372	0.3336	0.3300	0.3264	0.3228	0.3192	0.3156	0.3121
-0.3	0.3821	0.3783	0.3745	0.3707	0.3669	0.3632	0.3594	0.3557	0.3520	0.3483
-0.2	0.4207	0.4168	0.4129	0.4090	0.4052	0.4013	0.3974	0.3936	0.3897	0.3859
-0.1	0.4602	0.4562	0.4522	0.4483	0.4443	0.4404	0.4364	0.4325	0.4286	0.4247
-0.0	0.5000	0.4960	0.4920	0.4880	0.4840	0.4801	0.4761	0.4721	0.4681	0.4641
0.0	0.5000	0.5040	0.5080	0.5120	0.5160	0.5199	0.5239	0.5279	0.5319	0.5359
0.1	0.5398	0.5438	0.5478	0.5517	0.5557	0.5596	0.5636	0.5675	0.5714	0.5753
0.2	0.5793	0.5832	0.5871	0.5910	0.5948	0.5987	0.6026	0.6064	0.6103	0.6141
0.3	0.6179	0.6217	0.6255	0.6293	0.6331	0.6368	0.6406	0.6443	0.6480	0.6517
0.4	0.6554	0.6591	0.6628	0.6664	0.6700	0.6736	0.6772	0.6808	0.6844	0.6879
0.5	0.6915	0.6950	0.6985	0.7019	0.7054	0.7088	0.7123	0.7157	0.7190	0.7224
0.6	0.7257	0.7291	0.7324	0.7357	0.7389	0.7422	0.7454	0.7486	0.7517	0.7549
0.7	0.7580	0.7611	0.7642	0.7673	0.7704	0.7734	0.7764	0.7794	0.7823	0.7852
0.8	0.7881	0.7910	0.7939	0.7967	0.7995	0.8023	0.8051	0.8078	0.8106	0.8133
0.9	0.8159	0.8186	0.8212	0.8238	0.8264	0.8289	0.8315	0.8340	0.8365	0.8389
1.0	0.8413	0.8438	0.8461	0.8485	0.8508	0.8531	0.8554	0.8577	0.8599	0.8621
1.1	0.8643	0.8665	0.8686	0.8708	0.8729	0.8749	0.8770	0.8790	0.8810	0.8830
1.2	0.8849	0.8869	0.8888	0.8907	0.8925	0.8944	0.8962	0.8980	0.8997	0.9015
1.3	0.9032	0.9049	0.9066	0.9082	0.9099	0.9115	0.9131	0.9147	0.9162	0.9177
1.4	0.9192	0.9207	0.9222	0.9236	0.9251	0.9265	0.9278	0.9292	0.9306	0.9319
1.5	0.9332	0.9345	0.9357	0.9370	0.9382	0.9394	0.9406	0.9418	0.9429	0.9441
1.6	0.9452	0.9463	0.9474	0.9484	0.9495	0.9505	0.9515	0.9525	0.9535	0.9545
1.7	0.9554	0.9564	0.9573	0.9582	0.9591	0.9599	0.9608	0.9616	0.9625	0.9633
1.8	0.9641	0.9649	0.9656	0.9664	0.9671	0.9678	0.9686	0.9693	0.9699	0.9706
1.9	0.9713	0.9719	0.9726	0.9732	0.9738	0.9744	0.9750	0.9756	0.9761	0.9767
2.0	0.9772	0.9778	0.9783	0.9788	0.9793	0.9798	0.9803	0.9808	0.9812	0.9817
2.1	0.9821	0.9826	0.9830	0.9834	0.9838	0.9842	0.9846	0.9850	0.9854	0.9857
2.2	0.9861	0.9864	0.9868	0.9871	0.9875	0.9878	0.9881	0.9884	0.9887	0.9890
2.3	0.9893	0.9896	0.9898	0.9901	0.9904	0.9906	0.9909	0.9911	0.9913	0.9916
2.4	0.9918	0.9920	0.9922	0.9925	0.9927	0.9929	0.9931	0.9932	0.9934	0.9936
2.5	0.9938	0.9940	0.9941	0.9943	0.9945	0.9946	0.9948	0.9949	0.9951	0.9952
2.6	0.9953	0.9955	0.9956	0.9957	0.9959	0.9960	0.9961	0.9962	0.9963	0.9964
2.7	0.9965	0.9966	0.9967	0.9968	0.9969	0.9970	0.9971	0.9972	0.9973	0.9974
2.8	0.9974	0.9975	0.9976	0.9977	0.9977	0.9978	0.9979	0.9979	0.9980	0.9981
2.9	0.9981	0.9982	0.9982	0.9983	0.9984	0.9984	0.9985	0.9985	0.9986	0.9986
3.0	0.99865	0.99869	0.99874	0.99878	0.99882	0.99886	0.99889	0.99893	0.99896	0.99900
3.1	0.99903	0.99906	0.99910	0.99913	0.99915	0.99918	0.99921	0.99924	0.99926	0.99929
3.2	0.99931	0.99934	0.99936	0.99938	0.99940	0.99942	0.99944	0.99946	0.99948	0.99950
3.3	0.99952	0.99953	0.99955	0.99957	0.99958	0.99960	0.99961	0.99962	0.99964	0.99965
3.4	0.99966	0.99967	0.99969	0.99970	0.99971	0.99972	0.99973	0.99974	0.99975	0.99976

TABLE A.1 · *Areas Under the Normal Curve*

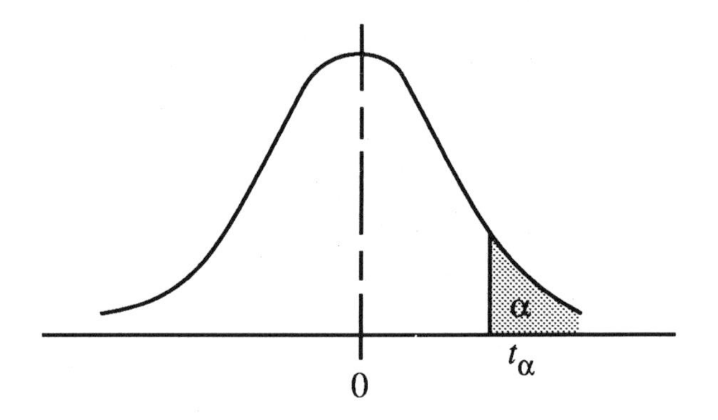

ν	α				
	0.10	0.05	0.025	0.01	0.005
1	3.078	6.314	12.706	31.821	63.657
2	1.886	2.920	4.303	6.965	9.925
3	1.638	2.353	3.182	4.541	5.841
4	1.533	2.132	2.776	3.747	4.604
5	1.476	2.015	2.571	3.365	4.032
6	1.440	1.943	2.447	3.143	3.707
7	1.415	1.895	2.365	2.998	3.499
8	1.397	1.860	2.306	2.896	3.355
9	1.383	1.833	2.262	2.821	3.250
10	1.372	1.812	2.228	2.764	3.169
11	1.363	1.796	2.201	2.718	3.106
12	1.356	1.782	2.179	2.681	3.055
13	1.350	1.771	2.160	2.650	3.012
14	1.345	1.761	2.145	2.624	2.977
15	1.341	1.753	2.131	2.602	2.947
16	1.337	1.746	2.120	2.583	2.921
17	1.333	1.740	2.110	2.567	2.898
18	1.330	1.734	2.101	2.552	2.878
19	1.328	1.729	2.093	2.539	2.861
20	1.325	1.725	2.086	2.528	2.845
21	1.323	1.721	2.080	2.518	2.831
22	1.321	1.717	2.074	2.508	2.819
23	1.319	1.714	2.069	2.500	2.807
24	1.318	1.711	2.064	2.492	2.797
25	1.316	1.708	2.060	2.485	2.787
26	1.315	1.706	2.056	2.479	2.779
27	1.314	1.703	2.052	2.473	2.771
28	1.313	1.701	2.048	2.467	2.763
29	1.311	1.699	2.045	2.462	2.756
inf.	1.282	1.645	1.960	2.326	2.576

TABLE A.2 · *Percentage points of the* t-*distribution*

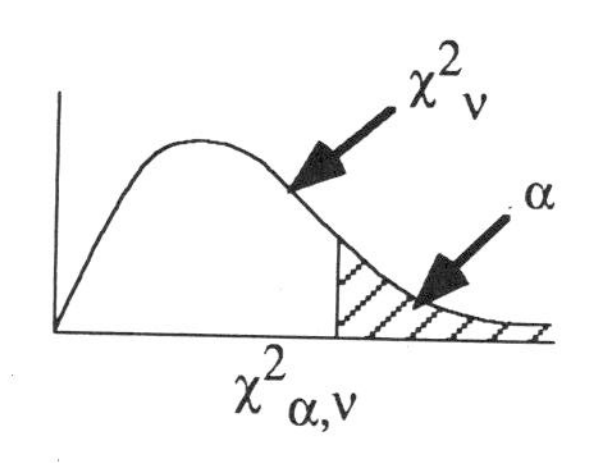

$\nu \backslash \alpha$	.995	.99	.975	.95	.90	.75	.5	.25	.10	.05	.025	.01	.005	.001
1					0.02	0.10	0.45	1.32	2.71	3.84	5.02	6.63	7.88	10.83
2	0.01	0.02	0.05	0.10	0.21	0.58	1.39	2.77	4.61	5.99	7.38	9.21	10.60	13.82
3	0.07	0.11	0.22	0.35	0.58	1.21	2.37	4.11	6.25	7.81	79.35	11.34	12.84	16.27
4	0.21	0.30	0.48	0.71	1.06	1.92	3.36	5.39	7.78	9.49	11.14	13.28	14.86	18.47
5	0.41	0.55	0.83	1.15	1.61	2.67	4.35	6.63	9.24	11.07	12.83	15.09	16.75	20.52
6	0.68	0.87	1.24	1.64	2.20	3.45	5.35	7.84	10.64	12.59	14.45	16.81	18.55	22.46
7	0.99	1.24	1.69	2.17	2.83	4.25	6.35	9.04	12.02	14.07	16.01	18.48	20.28	24.32
8	1.34	1.65	2.18	2.73	3.49	5.07	7.34	10.22	13.36	15.51	17.53	20.09	21.96	26.12
9	1.73	2.09	2.70	3.33	4.17	5.90	8.34	11.39	14.68	16.92	19.02	21.67	23.59	27.88
10	2.16	2.56	3.25	3.94	4.87	6.74	9.34	12.55	15.99	18.31	20.48	23.21	25.19	29.59
11	2.60	3.05	3.82	4.57	5.58	7.58	10.34	13.70	17.28	19.68	21.92	24.72	26.76	31.26
12	3.07	3.57	4.40	5.23	6.30	8.44	11.34	14.85	18.55	21.03	23.34	26.22	28.30	32.91
13	3.57	4.11	5.01	5.89	7.04	9.30	12.34	15.98	19.81	22.36	24.74	27.69	29.82	34.53
14	4.07	4.66	5.63	6.57	7.79	10.17	13.34	17.12	21.06	23.68	26.12	29.14	31.32	36.12
15	4.60	5.23	6.26	7.26	8.55	11.04	14.34	18.25	22.31	25.00	27.49	30.58	32.80	37.70
16	5.14	5.81	6.91	7.96	9.31	11.91	15.34	19.37	23.54	26.30	28.85	32.00	34.27	39.25
17	5.70	6.41	7.56	8.67	10.09	12.79	16.34	20.49	24.77	27.59	30.19	33.41	35.73	40.79
18	6.26	7.01	8.23	9.39	10.86	13.68	17.34	21.60	25.99	28.87	31.53	34.81	37.16	42.31
19	6.84	7.63	8.91	10.12	11.65	14.56	18.34	22.72	27.20	30.14	32.85	36.19	38.58	43.82
20	7.43	8.26	9.59	10.85	12.44	15.45	19.34	23.83	28.41	31.41	34.17	37.57	40.00	45.32
21	8.03	8.90	10.28	11.59	13.24	16.34	20.34	24.93	29.62	32.67	35.48	38.93	41.40	46.80
22	8.64	9.54	10.98	12.34	14.04	17.24	21.34	26.04	30.81	33.92	36.78	40.29	42.80	48.27
23	9.26	10.20	11.69	13.09	14.85	18.14	22.34	27.14	32.01	35.17	38.08	41.64	44.18	49.73
24	9.89	10.86	12.40	13.85	15.66	19.04	23.34	28.24	33.20	36.42	39.36	42.98	45.56	51.18
25	10.52	11.52	13.12	14.61	16.47	19.94	24.34	29.34	34.38	37.65	40.65	44.31	46.93	52.62
30	13.79	14.95	16.79	18.49	20.60	24.48	29.34	34.80	40.26	43.77	46.98	50.89	53.67	59.70
40	20.71	22.16	24.43	26.51	29.05	33.66	39.34	45.62	51.80	55.76	59.34	63.69	66.77	73.40
50	27.99	29.71	32.36	34.76	37.69	42.94	49.33	56.33	63.17	67.50	71.42	76.15	79.49	86.66
60	35.53	37.48	40.48	43.19	46.46	52.29	59.33	66.98	74.40	79.08	83.30	88.38	91.95	99.61
70	43.28	45.44	48.76	51.74	55.33	61.70	69.33	77.58	85.53	90.53	95.02	100.42	104.22	112.32
80	51.17	53.54	57.15	60.39	64.28	71.14	79.33	88.13	96.58	101.88	106.63	112.33	116.32	124.84
90	59.20	61.75	65.65	69.13	73.29	80.62	89.33	98.64	107.56	113.14	118.14	124.12	128.30	137.21
100	67.33	70.06	74.22	77.93	82.36	90.13	99.33	109.14	118.50	124.34	129.56	135.81	140.17	149.45

TABLE A.3 · *Percentage Points of the χ^2 Distribution*

c/λ	.01	.05	.10	.20	.30	.40	.50	.60	.70	.80	.90	1.00
0	.990	.951	.904	.818	.740	.670	.606	.548	.496	.449	.406	.367
1	.999	.998	.995	.982	.963	.938	.909	.878	.844	.808	.772	.735
2		.999	.999	.998	.996	.992	.985	.976	.965	.952	.937	.919
3				.999	.999	.999	.998	.996	.994	.990	.986	.981
4							.999	.999	.999	.998	.997	.996
5										.999	.999	.999

c/λ	1.10	1.20	1.30	1.40	1.50	1.60	1.70	1.80	1.90	2.00	2.10	2.20
0	.332	.301	.272	.246	.223	.201	.182	.165	.149	.135	.122	.110
1	.699	.662	.626	.591	.557	.524	.493	.462	.433	.406	.379	.354
2	.900	.879	.857	.833	.808	.783	.757	.730	.703	.676	.649	.622
3	.974	.966	.956	.946	.934	.921	.906	.891	.874	.857	.838	.819
4	.994	.992	.989	.985	.981	.976	.970	.963	.955	.947	.937	.927
5	.999	.998	.997	.996	.995	.993	.992	.989	.986	.983	.979	.975
6		.999	.999	.999	.999	.998	.998	.997	.996	.995	.994	.992
7						.999	.999	.999	.999	.998	.998	.998
8										.999	.999	.999

c/λ	2.30	2.40	2.50	2.60	2.70	2.80	2.90	3.00	3.50	4.00	4.50	5.00
0	.100	.090	.082	.074	.067	.060	.055	.049	.030	.018	.011	.006
1	.330	.308	.287	.267	.248	.231	.214	.199	.135	.091	.061	.040
2	.596	.569	.543	.518	.493	.469	.445	.423	.320	.238	.173	.124
3	.799	.778	.757	.736	.714	.691	.669	.647	.536	.433	.342	.265
4	.916	.904	.891	.877	.862	.847	.831	.815	.725	.628	.532	.440
5	.970	.964	.957	.950	.943	.934	.925	.916	.857	.785	.702	.615
6	.990	.988	.985	.982	.979	.975	.971	.966	.934	.889	.831	.762
7	.997	.996	.995	.994	.993	.991	.990	.988	.973	.948	.913	.866
8	.999	.999	.998	.998	.998	.997	.996	.996	.990	.978	.959	.931
9			.999	.999	.999	.999	.999	.998	.996	.991	.982	.968
10								.999	.998	.997	.993	.986
11									.999	.999	.997	.994
12											.999	.997
13												.999

TABLE A.4 · *Cumulative Poisson Probabilities – continued*

c/λ	5.50	6.00	6.50	7.00	7.50	8.00	8.50	9.00	9.50	10.0	15.0	20.0
0	.004	.002	.001	.000	.000	.000	.000	.000	.000	.000	.000	.000
1	.026	.017	.011	.007	.004	.003	.001	.001	.000	.000	.000	.000
2	.088	.061	.043	.029	.020	.013	.009	.006	.004	.002	.000	.000
3	.201	.151	.111	.081	.059	.042	.030	.021	.014	.010	.000	.000
4	.357	.285	.223	.172	.132	.099	.074	.054	.040	.029	.000	.000
5	.528	.445	.369	.300	.241	.191	.149	.115	.088	.067	.002	.000
6	.686	.606	.526	.449	.378	.313	.256	.206	.164	.130	.007	.000
7	.809	.743	.672	.598	.524	.452	.385	.323	.268	.220	.018	.000
8	.894	.847	.791	.729	.661	.592	.523	.455	.391	.332	.037	.002
9	.946	.916	.877	.830	.776	.716	.652	.587	.521	.457	.069	.005
10	.974	.957	.933	.901	.862	.815	.763	.705	.645	.583	.118	.070
11	.989	.979	.966	.946	.920	.888	.848	.803	.751	.696	.184	.021
12	.995	.991	.983	.973	.957	.936	.909	.875	.836	.791	.267	.039
13	.998	.996	.992	.987	.978	.965	.948	.926	.898	.864	.363	.066
14	.999	.998	.997	.994	.989	.982	.972	.958	.940	.916	.465	.104
15		.999	.998	.997	.995	.991	.986	.997	.966	.951	.568	.156
16			.999	.999	.998	.996	.993	.988	.982	.974	.664	.221
17					.999	.998	.997	.994	.991	.985	.748	.297
18						.999	.998	.997	.995	.992	.819	.381
19							.999	.998	.998	.996	.875	.470
20								.999	.999	.998	.917	.559
21										.999	.946	.643
22											.967	.720
23											.980	.787
24											.988	.843
25											.993	.887
26											.996	.922
27											.998	.947
28											.999	.965
29												.978
30												.986
31												.991
32												.995
33												.997
34												.998

TABLE A.4 · *Cumulative Poisson Probabilities (continued)*

TOTAL NUMBER OF FAILURES	CONFIDENCE INTERVALS					
	40 PERCENT		60 PERCENT		80 PERCENT	
	70 PERCENT LOWER LIMIT	70 PERCENT UPPER LIMIT	80 PERCENT LOWER LIMIT	80 PERCENT UPPER LIMIT	90 PERCENT LOWER LIMIT	90 PERCENT UPPER LIMIT
1	0.801	2.804	0.621	4.481	0.434	9.491
2	0.820	1.823	0.668	2.426	0.514	3.761
3	0.830	1.568	0.701	1.954	0.564	2.722
4	0.840	1.447	0.725	1.742	0.599	2.293
5	0.849	1.376	0.744	1.618	0.626	2.055
6	0.856	1.328	0.759	1.537	0.647	1.904
7	0.863	1.294	0.771	1.479	0.665	1.797
8	0.869	1.267	0.782	1.435	0.680	1.718
9	0.874	1.247	0.796	1.400	0.693	1.657
10	0.878	1.230	0.799	1.372	0.704	1.607
11	0.882	1.215	0.806	1.349	0.714	1.567
12	0.886	1.203	0.812	1.329	0.723	1.533
13	0.889	1.193	0.818	1.312	0.731	1.504
14	0.892	1.184	0.823	1.297	0.738	1.478
15	0.895	1.176	0.828	1.284	0.745	1.456
16	0.897	1.169	0.832	1.272	0.751	1.437
17	0.900	1.163	0.836	1.262	0.757	1.419
18	0.902	1.157	0.840	1.253	0.763	1.404
19	0.904	1.152	0.843	1.244	0.767	1.390
20	0.906	1.147	0.846	1.237	0.772	1.377
30	0.920	1.115	0.870	1.185	0.806	1.291

TABLE A.5 · *Demonstrated MTBF confidence limit multipliers, for failure calculation (MIL-HDBK-781 Table 13)*

TOTAL NUMBER OF FAILURES	CONFIDENCE INTERVALS					
	40 PERCENT		60 PERCENT		80 PERCENT	
	70 PERCENT LOWER LIMIT	70 PERCENT UPPER LIMIT	80 PERCENT LOWER LIMIT	80 PERCENT UPPER LIMIT	90 PERCENT LOWER LIMIT	90 PERCENT UPPER LIMIT
1	0.410	2.804	0.334	4.481	0.257	9.491
2	0.533	1.823	0.467	2.426	0.376	3.761
3	0.630	1.568	0.544	1.954	0.449	2.722
4	0.679	1.447	0.595	1.742	0.500	2.293
5	0.714	1.376	0.632	1.618	0.539	2.055
6	0.740	1.328	0.661	1.537	0.570	1.904
7	0.760	1.294	0.684	1.479	0.595	1.797
8	0.777	1.267	0.703	1.435	0.616	1.718
9	0.790	1.247	0.719	1.400	0.634	1.657
10	0.802	1.230	0.733	1.372	0.649	1.607
11	0.123	1.215	0.744	1.349	0.663	1.567
12	0.821	1.203	0.755	1.329	0.675	1.533
13	0.828	1.193	0.764	1.312	0.686	1.504
14	0.835	1.184	0.772	1.297	0.696	1.478
15	0.841	1.176	0.780	1.284	0.705	1.456
16	0.847	1.169	0.787	1.272	0.713	1.437
17	0.852	1.163	0.793	1.262	0.720	1.419
18	0.856	1.157	0.799	1.253	0.727	1.404
19	0.861	1.152	0.804	1.224	0.734	1.390
20	0.864	1.147	0.809	1.237	0.740	1.377
30	0.891	1.115	0.844	1.185	0.783	1.291

TABLE A.6 · *Demonstrated MTBF confidence limit multipliers, for time calculation (MIL-HDBK-781 Table 14)*

Test Plan	Accept Times [1]				
IX-D	$T_{0} = 4.2$	$T_{1} = 6.1$	$T_{2} = 7.9$	$T_{3} = 9.4$	$T_{4} = 11.0$
	$T_{5} = 12.4$	$T_{6} = 13.9$	$T_{7} = 15.3$	$T_{8} = 16.6$	$T_{9} = 18.0$
	$T_{10} = 19.3$	$T_{11} = 20.7$	$T_{12} = 22.0$	$T_{13} = 23.3$	$T_{14} = 24.5$
	$T_{15} = 25.8$	$T_{11} = 27.1$	$T_{17} = 28.3$	$T_{18} = 29.6$	$T_{19} = 30.8$
	$T_{20} = 32.1$	$T_{21} = 33.3$	$T_{22} = 34.5$	$T_{23} = 35.8$	$T_{24} = 37.0$
	$T_{25} = 38.2$	$T_{26} = 39.4$	$T_{27} = 40.6$	$T_{28} = 41.8$	$T_{29} = 43.0$
	$T_{30} = 44.2$	$T_{31} = 45.4$	$T_{32} = 46.6$	$T_{33} = 47.8$	$T_{34} = 49.0$
	$T_{35} = 50.1$	$T_{36} = 51.3$	$T_{12} = 52.5$	$T_{38} = 53.7$	$T_{39} = 54.8$
	$T_{40} = 56.0$	$T_{41} = 57.2$	$T_{42} = 58.3$	$T_{43} = 59.5$	$T_{44} = 60.7$
	$T_{45} = 61.8$	$T_{46} = 63.0$	$T_{47} = 64.1$	$T_{48} = 65.3$	$T_{49} = 66.5$
	$T_{50} = 67.6$	$T_{51} = 68.8$	$T_{52} = 69.9$	$T_{53} = 71.1$	$T_{54} = 72.2$
X-D	$T_{0} = 3.2$	$T_{1} = 5.0$	$T_{2} = 6.6$	$T_{3} = 8.1$	$T_{4} = 9.5$
	$T_{5} = 10.9$	$T_{6} = 12.2$	$T_{7} = 13.6$	$T_{8} = 14.9$	$T_{9} = 16.1$
	$T_{10} = 17.4$	$T_{11} = 18.7$	$T_{12} = 19.9$	$T_{13} = 21.2$	$T_{14} = 22.4$
	$T_{15} = 23.6$	$T_{16} = 24.8$	$T_{17} = 26.1$	$T_{18} = 27.3$	$T_{19} = 28.4$
	$T_{20} = 29.6$	$T_{21} = 30.8$	$T_{22} = 32.0$	$T_{23} = 33.2$	$T_{24} = 34.4$
	$T_{25} = 35.6$	$T_{26} = 36.7$	$T_{27} = 37.9$	$T_{28} = 39.1$	$T_{29} = 40.2$
	$T_{30} = 41.4$	$T_{31} = 42.5$	$T_{32} = 43.7$	$T_{33} = 44.8$	$T_{34} = 46.0$
	$T_{35} = 47.1$	$T_{36} = 48.3$	$T_{37} = 49.4$	$T_{38} = 50.6$	$T_{39} = 51.7$
XI-D	$T_{0} = 3.0$	$T_{1} = 4.8$	$T_{2} = 6.3$	$T_{3} = 7.8$	$T_{4} = 9.2$
	$T_{5} = 10.5$	$T_{6} = 11.9$	$T_{7} = 13.2$	$T_{8} = 14.4$	$T_{9} = 15.7$
	$T_{10} = 17.0$	$T_{11} = 18.2$	$T_{12} = 19.5$	$T_{13} = 20.7$	$T_{14} = 21.9$
	$T_{15} = 23.1$	$T_{16} = 24.3$	$T_{17} = 25.5$	$T_{18} = 26.7$	$T_{19} = 27.9$
	$T_{20} = 29.1$	$T_{21} = 30.3$	$T_{22} = 31.4$	$T_{23} = 32.6$	

[1] Accept at time (*Tj*) if (*j*) failures have occurred up to that time.

Table A.7 · *Accept times of fixed-duration test plans, Program Manager's assessment – continued (MIL-HDBK-781 Table 15)*

Test plan	Accept Times [1]				
XII-D	T_0 = 3.7	T_1 = 5.6	T_2 = 7.2	T_3 = 8.8	T_4 = 10.3
	T_5 = 11.7	T_6 = 13.1	T_7 = 14.4	T_8 = 15.8	T_9 = 17.1
	T_{10} = 18.4	T_{11} = 19.7	T_{12} = 21.0	T_{13} = 22.3	T_{14} = 23.5
	T_{15} = 24.8	T_{16} = 26.0			
XIII-D	T_0 = 2.8	T_1 = 4.6	T_2 = 6.1	T_3 = 7.5	T_4 = 8.9
	T_5 = 10.3	T_6 = 11.6	T_7 = 12.9	T_8 = 14.1	T_9 = 15.4
	T_{10} = 16.6	T_{11} = 17.9	T_{12} = 19.1		
XIV-D	T_0 = 2.7	T_1 = 4.4	T_2 = 5.9	T_3 = 7.3	T_4 = 8.7
	T_5 = 10.0	T_6 = 11.3	T_7 = 12.6		
XV-D	T_0 = 3.5	T_1 = 5.4	T_2 = 7.0	T_3 = 8.6	T_4 = 10.0
	T_5 = 11.4	T_6 = 12.8			
XVI-D	T_0 = 2.5	T_1 = 4.1	T_2 = 5.6	T_3 = 7.0	T_4 = 8.3
XVII-D	T_0 = 2.2	T_1 = 3.8	T_2 = 5.2		
XIX-D	T_0 = 2.1	T_1 = 3.7	T_2 = 5.1	T_3 = 6.4	T_4 = 7.7
	T_5 = 8.9	T_6 = 10.2	T_7 = 11.4	T_8 = 12.6	
XX-D	T_0 = 1.8	T_1 = 3.2	T_2 = 4.5		
XXI-D	T_0 = 1.1				

[1] Accept at time (*Tj*) if (*j*) failures have occurred up to that time.

Table A.7 · *Accept times of fixed-duration test plans, Program Manager's assessment. (Continued) (MIL-HDBK-781 Table 15)*

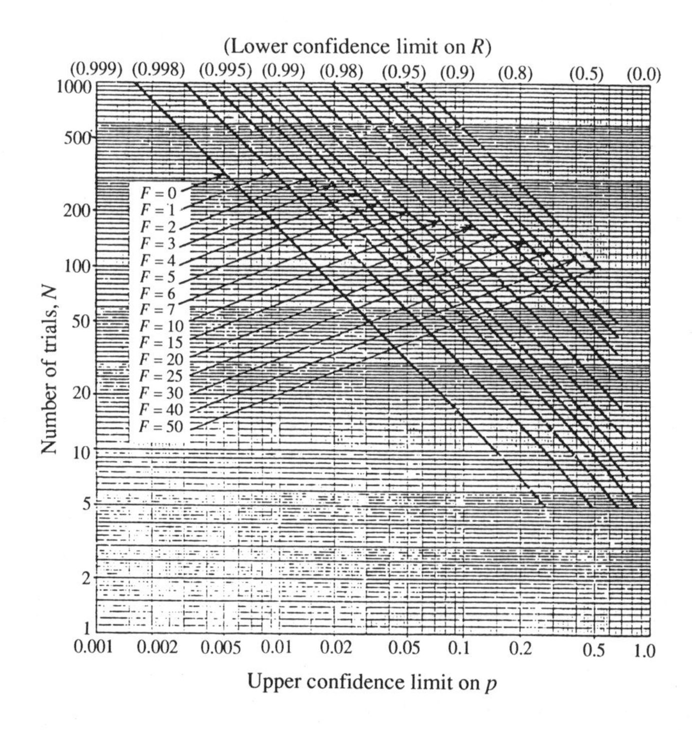

FIGURE A.1(a) · *Upper confidence limit on unreliability (one minus lower confidence limit on reliability); number of trials* N, *observed failures* F, *confidence coefficient* $\gamma = 0.80$. *Reproduced with permission from Lloyd, D. K. and Lipow, M.,* Reliability: Management, Methods and Mathematics, *2nd ed., Milwaukee, WI: ASQC, 1984*

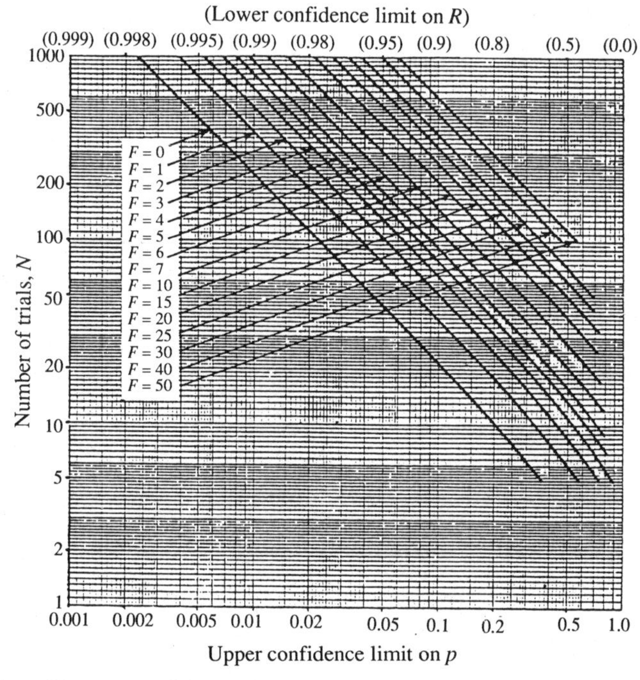

FIGURE A.1(b) · *Upper confidence limit on unreliability (one minus lower confidence limit on reliability); number of trials* N, *observed failures* F, *confidence coefficient* $\gamma = 0.90$. *Reproduced with permission from Lloyd, D. K. and Lipow, M.,* Reliability: Management, Methods and Mathematics, *2nd ed., Milwaukee, WI: ASQC, 1984*

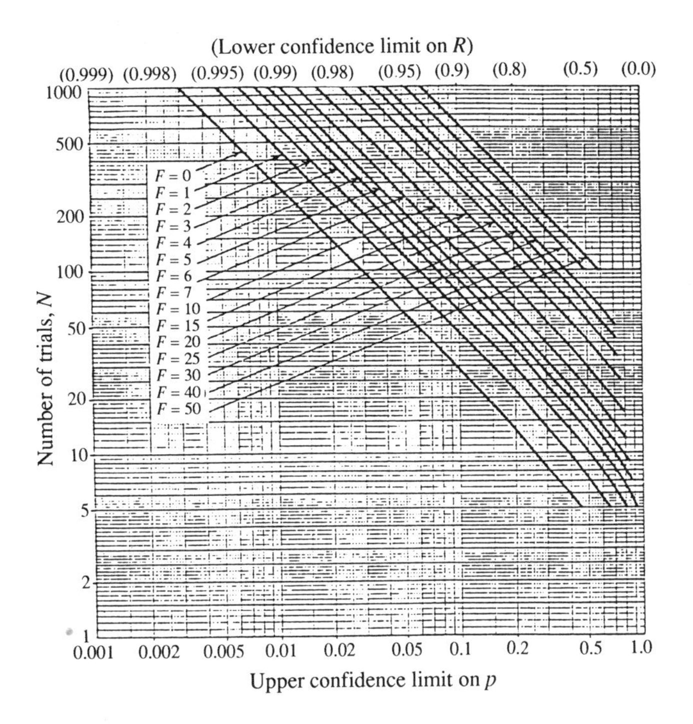

FIGURE A.1(c) · *Upper confidence limit on unreliability (one minus lower confidence limit on reliability); number of trials* N, *observed failures* F, *confidence coefficient* $\gamma = 0.95$. *Reproduced with permission from Lloyd, D. K. and Lipow, M.,* Reliability: Management, Methods and Mathematics, *2nd ed., Milwaukee, WI: ASQC, 1984*

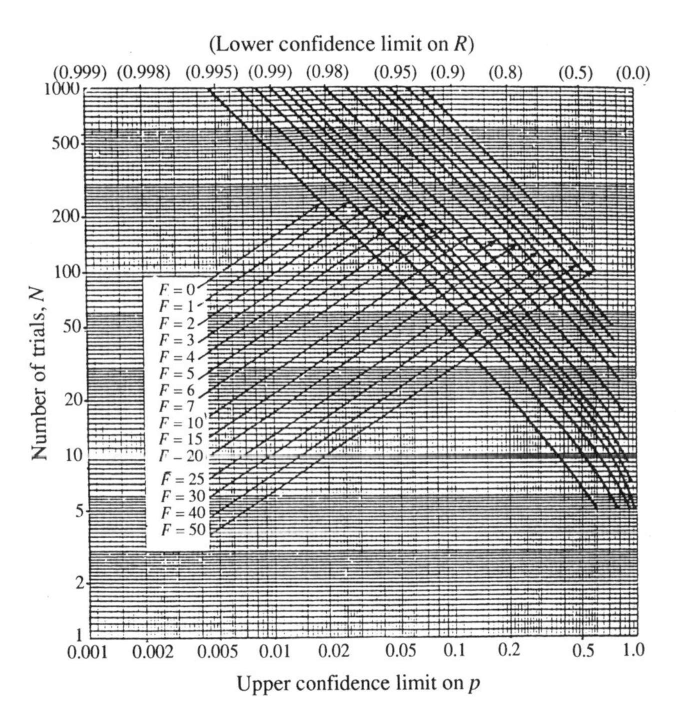

FIGURE A.1(d) · *Upper confidence limit on unreliability (one minus lower confidence limit on reliability); number of trials* N, *observed failures* F, *confidence coefficient* $\gamma = 0.99$. *Reproduced with permission from Lloyd, D. K. and Lipow, M.,* Reliability: Management, Methods and Mathematics, *2nd ed., Milwaukee, WI: ASQC, 1984*

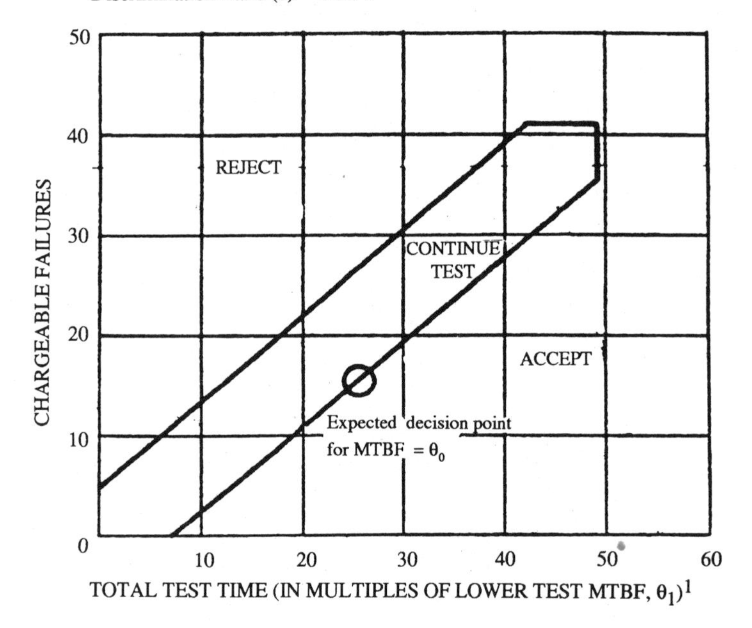

CHARGEABLE FAILURES	STANDARDIZED TERMINATION TIME, t [2]		CHARGEABLE FAILURES	STANDARDIZED TERMINATION TIME, t [2]	
	REJECT AT $t_R \leq$	ACCEPT AT $t_A \geq$		REJECT AT $t_R \leq$	ACCEPT AT $t_A \geq$
0	N/A	6.95	21	18.50	32.49
1	N/A	8.17	22	19.80	33.70
2	N/A	9.38	23	21.02	34.92
3	N/A	10.60	24	22.23	36.13
4	N/A	11.80	25	23.45	37.35
5	N/A	13.03	26	24.66	38.57
6	0.34	14.25	27	25.88	39.78
7	1.56	15.46	28	27.07	41.00
8	2.78	16.69	29	28.31	42.22
9	3.99	17.90	30	29.53	43.43
10	5.20	19.11	31	30.74	44.65
11	6.42	20.33	32	31.96	45.86
12	7.64	21.54	33	33.18	47.08
13	8.86	22.76	34	34.39	48.30
14	10.07	23.98	35	35.61	49.50
15	11.29	25.19	36	36.82	49.50
16	12.50	26.41	37	38.04	49.50
17	13.72	27.62	38	39.26	49.50
18	14.94	28.64	39	40.47	49.50
19	16.15	30.06	40	41.69	49.50
20	17.37	31.27	41	49.50	N/A

Accept-reject criteria

[1]Total test time is the summation of operating time of all units included in test sample.

[2]To determine the actual termination time, multiply the standardized termination time (t) by the lower test MTBF (Θ_1).

FIGURE A.2(a) · *Test Plan I-D (MIL-HDBK-781 Figure 9)*

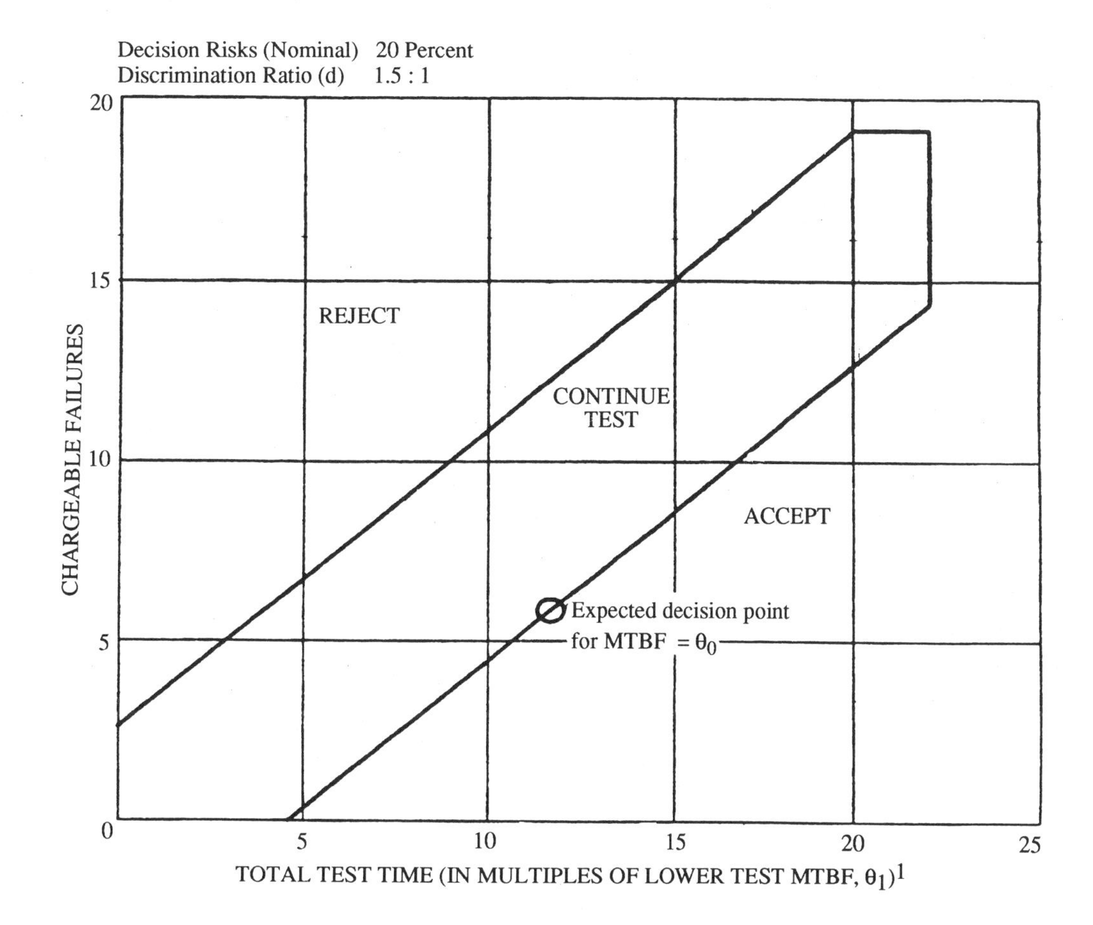

Chargeable Failures	Standardized Termination Time, t^2		Chargeable Failures	Standardized Termination Time, t^2	
	Reject at $t_R \leq$	Accept at $t_A \geq$		Reject at $t_R \leq$	Accept at $t_A \geq$
0	N/A	4.19	10	8.76	16.35
1	N/A	5.40	11	9.98	17.57
2	N/A	6.62	12	11.19	18.73
3	.24	7.83	13	12.41	19.99
4	1.46	9.05	14	13.62	21.21
5	2.67	10.26	15	14.84	22.35
6	3.90	11.49	16	16.05	21.90
7	5.12	12.71	17	17.28	21.90
8	6.33	13.92	18	18.50	21.90
9	7.55	15.14	19	21.90	N/A

Accept-reject criteria

[1]Total test time is the summation of operating time of all units included in test sample.

[2]To determine the actual termination time, multiply the standardized termination time (t) by the lower test MTBF (Θ_1).

FIGURE A.2(b) · *Test Plan II-D (MIL-HDBK-781 Figure 10)*

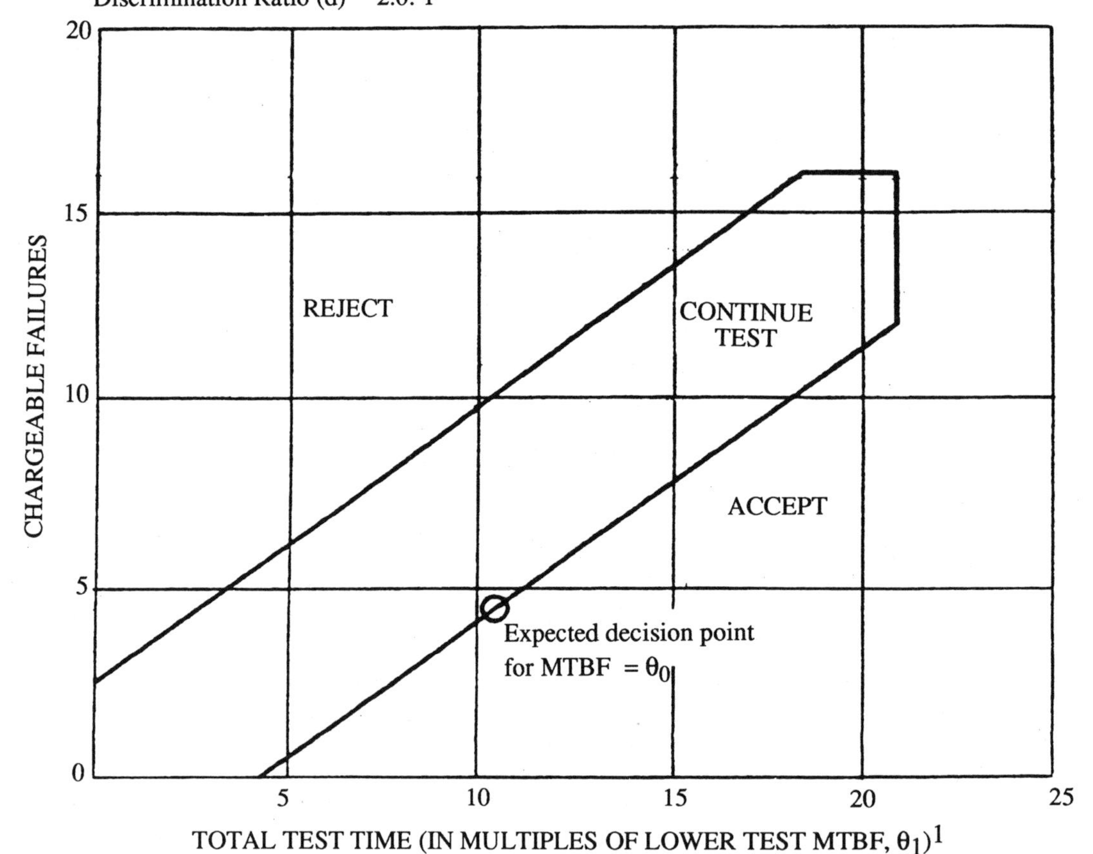

Chargeable Failures	Standardized Termination Time, t^2		Chargeable Failures	Standardized Termination Time, t^2	
	Reject at $t_R \leq$	Accept at $t_A \geq$		Reject at $t_R \leq$	Accept at $t_A \geq$
0	N/A	4.40	9	9.02	16.88
1	N/A	5.79	10	10.40	18.26
2	N/A	7.18	11	11.79	19.65
3	.70	8.56	12	13.81	20.60
4	2.08	9.94	13	14.56	20.60
5	3.48	11.34	14	15.94	20.60
6	4.86	12.72	15	17.34	20.60
7	6.24	14.10	16	20.60	N/A
8	7.63	15.49			

Accept-reject criteria

[1]Total test time is the summation of operating time of all units included in test sample.

[2]To determine the actual termination time, multiply the standardized termination time (t) by the lower test MTBF (θ_1).

FIGURE A.2(c) · *Test Plan III-D (MIL-HDBK-781 Figure 11)*

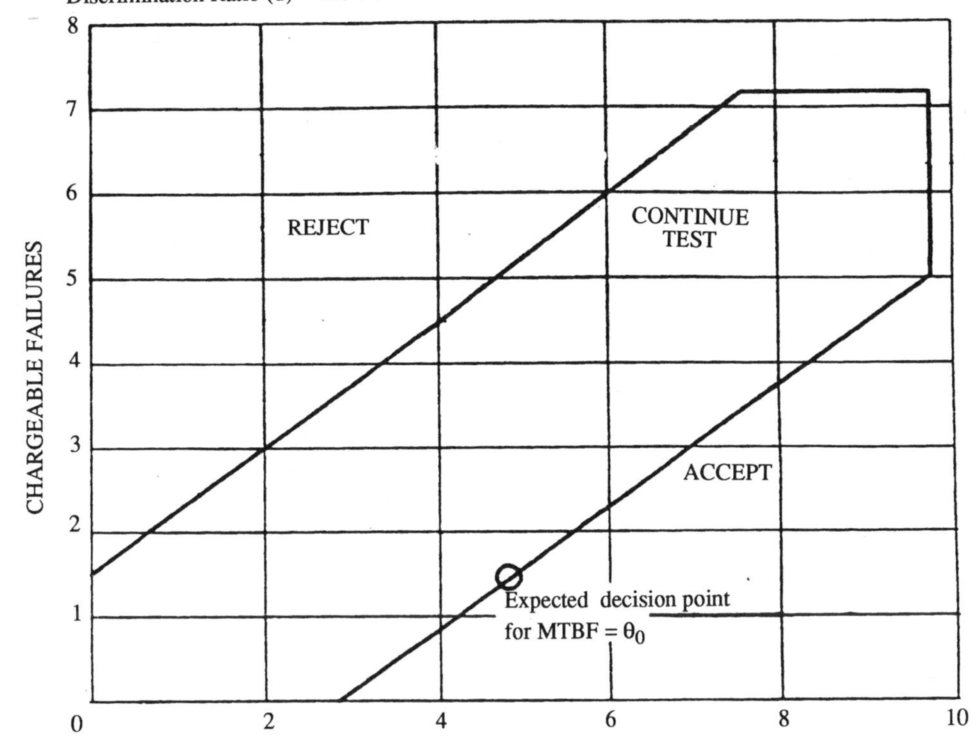

CHARGEABLE FAILURES	STANDARDIZED TERMINATION TIME, t[2]	
	REJECT AT $t_R \leq$	ACCEPT AT $t_A \geq$
0	N/A	2.80
1	N/A	4.18
2	.70	5.58
3	2.08	6.96
4	3.46	8.34
5	4.86	9.74
6	6.24	9.74
7	7.62	9.74
8	9.74	N/A

Accept-reject criteria

[1]Total test time is the summation of operating time of all units included in test sample.

[2]To determine the actual termination time, multiply the standardized termination time (t) by the lower test MTBF (Θ_1).

FIGURE A.2(d) · *Test Plan IV-D (MIL-HDBK-781 Figure 12)*

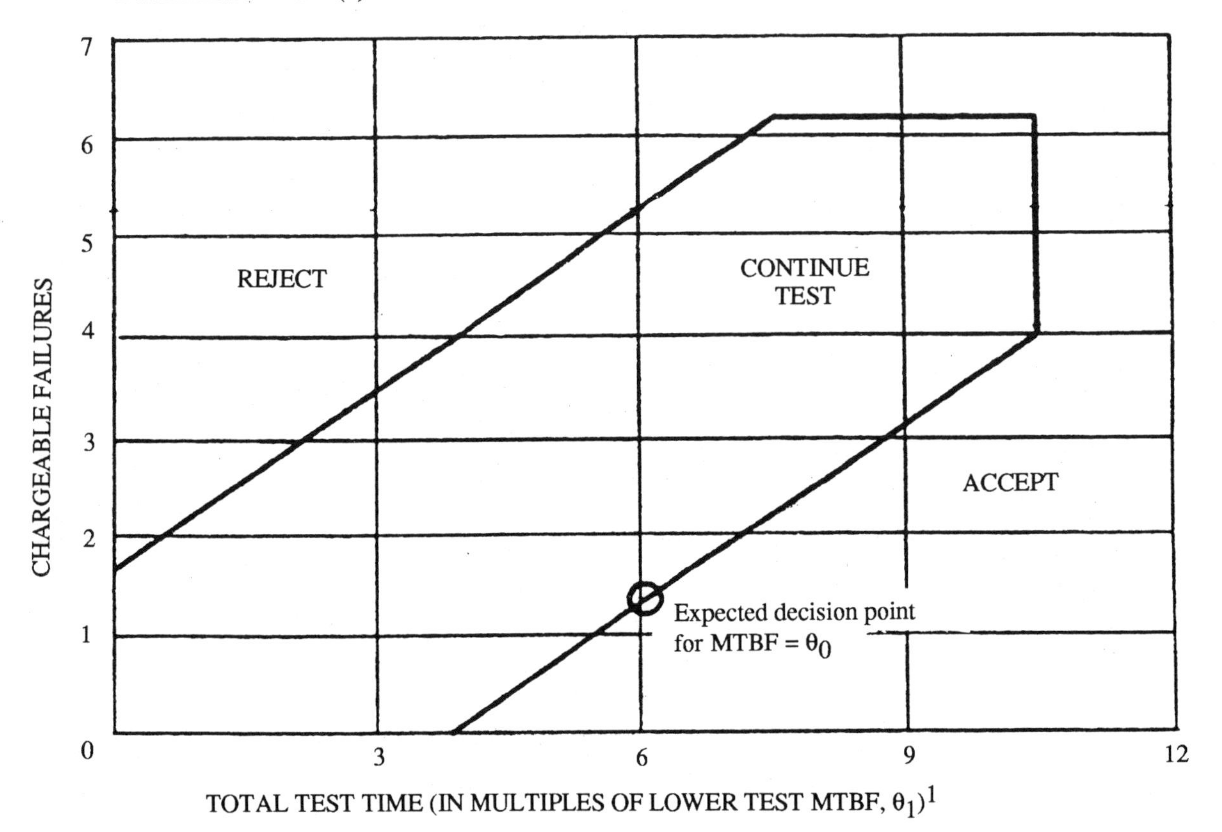

Chargeable Failures	Standardized Termination Time, t^2	
	Reject at $t_R \leq$	Accept at $t_A \geq$
0	N/A	3.75
1	N/A	5.40
2	.57	7.05
3	2.22	8.70
4	3.87	10.35
5	5.52	10.35
6	7.17	10.35
7	10.35	N/A

Accept-reject criteria

[1] Total test time is the summation of operating time of all units included in test sample.
[2] To determine the actual termination time, multiply the standardized termination time (t) by the lower test MTBF (θ_1).

FIGURE A.2(e) · *Test Plan V-D (MIL-HDBK-781 Figure 13)*

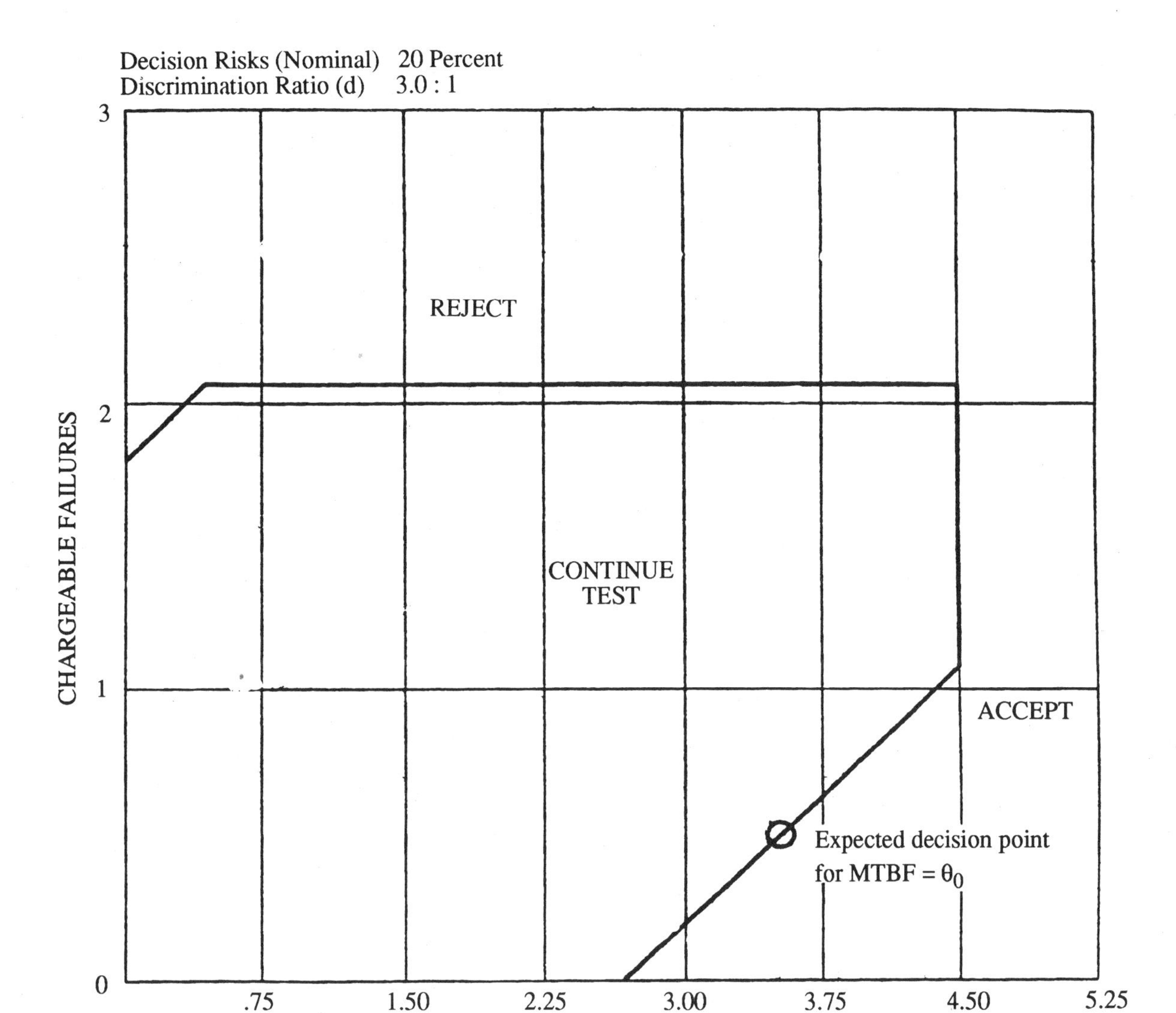

Chargeable Failures	Standardized Termination Time, t^2	
	Reject at $t_R \leq$	Accept at $t_A \geq$
0	N/A	2.67
1	N/A	4.32
2	.36	4.50
3	4.50	N/A

Accept-reject criteria

[1]Total test time is the summation of operating time of all units included in test sample.
[2]To determine the actual termination time, multiply the standardized termination time (t) by the lower test MTBF (θ_1).

FIGURE A.2(f) · *Test Plan VI-D (MIL-HDBK-781 Figure 14)*

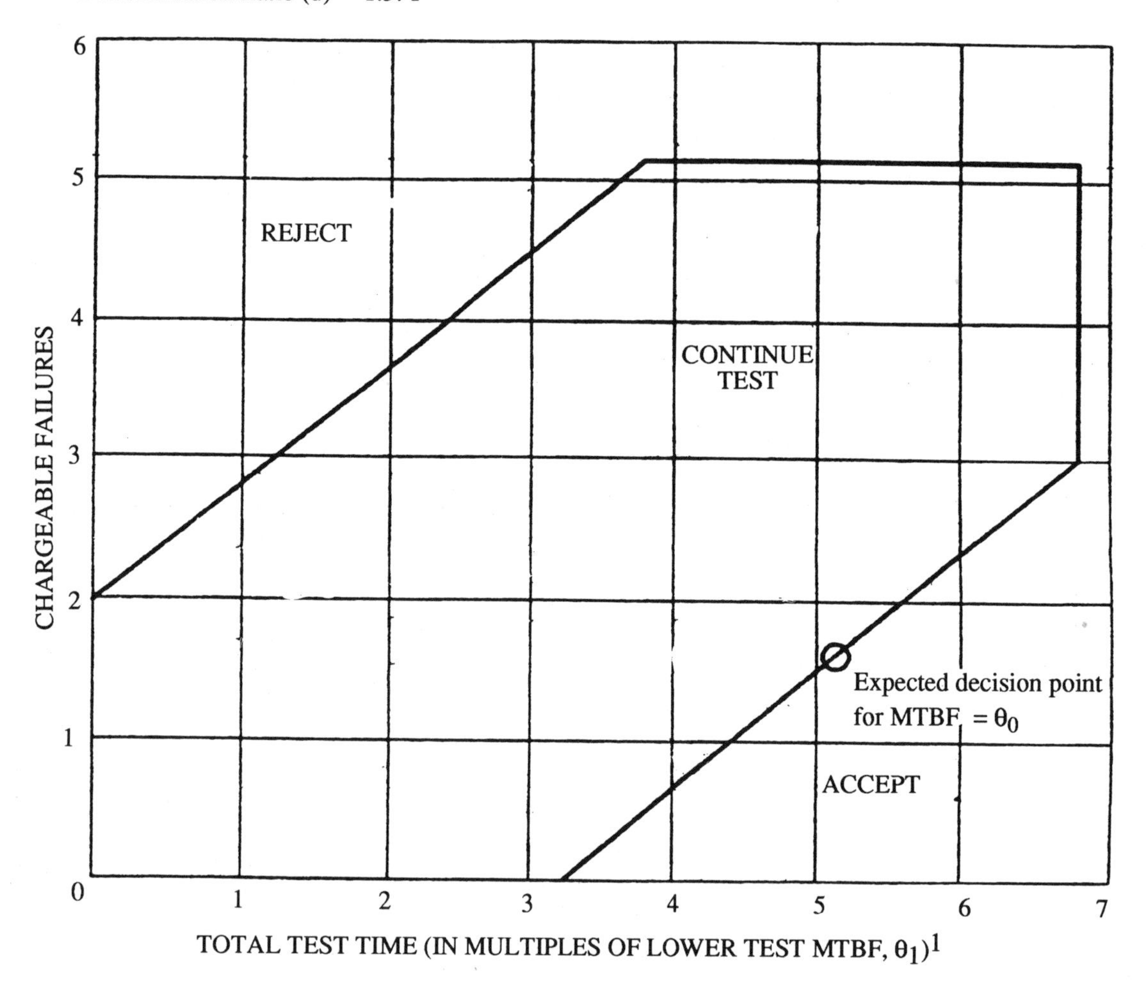

CHARGEABLE FAILURES	STANDARDIZED TERMINATION TIME, t[2]	
	REJECT AT $t_R \leq$	ACCEPT AT $t_A \geq$
0	N/A	3.15
1	N/A	4.37
2	N/A	5.58
3	1.22	6.80
4	2.43	6.80
5	3.65	6.80
6	6.80	N/A

Accept-reject criteria

[1]Total test time is the summation of operating time of all units included in test sample.
[2]To determine the actual termination time, multiply the standardized termination time (t) by the lower test MTBF (θ_1).

FIGURE A.2(g) · *Test Plan VII-D (MIL-HDBK-781 Figure 15)*

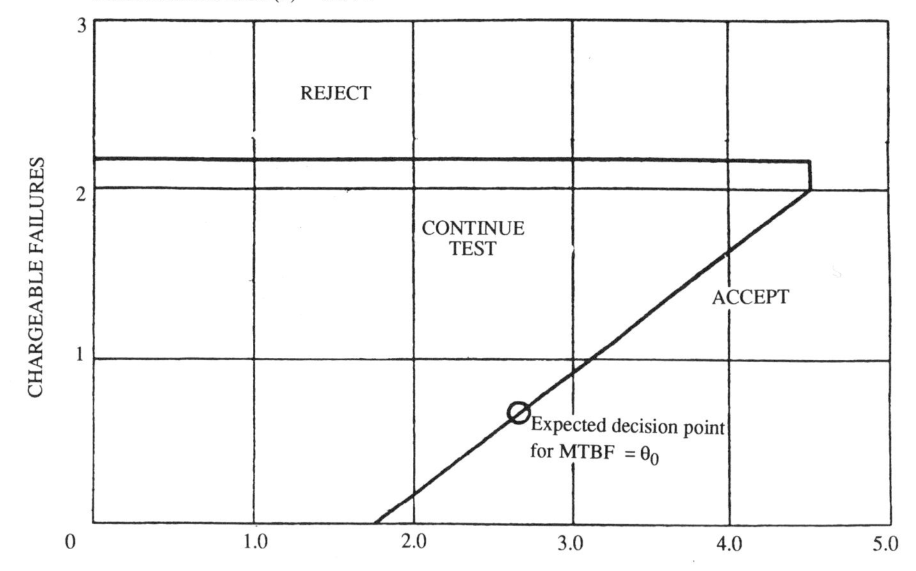

CHARGEABLE FAILURES	STANDARDIZED TERMINATION TIME, t^2	
	REJECT AT $t_R \leq$	ACCEPT AT $t_A \geq$
0	N/A	1.72
1	N/A	3.10
2	N/A	4.50
3	4.50	N/A

Accept-reject criteria

[1]Total test time is the summation of operating time of all units included in test sample.

[2]To determine the actual termination time, multiply the standardized termination time (t) by the lower test MTBF (θ_1).

FIGURE A.2(h) · *Test Plan VIII-D (MIL-HDBK-781 Figure 16)*

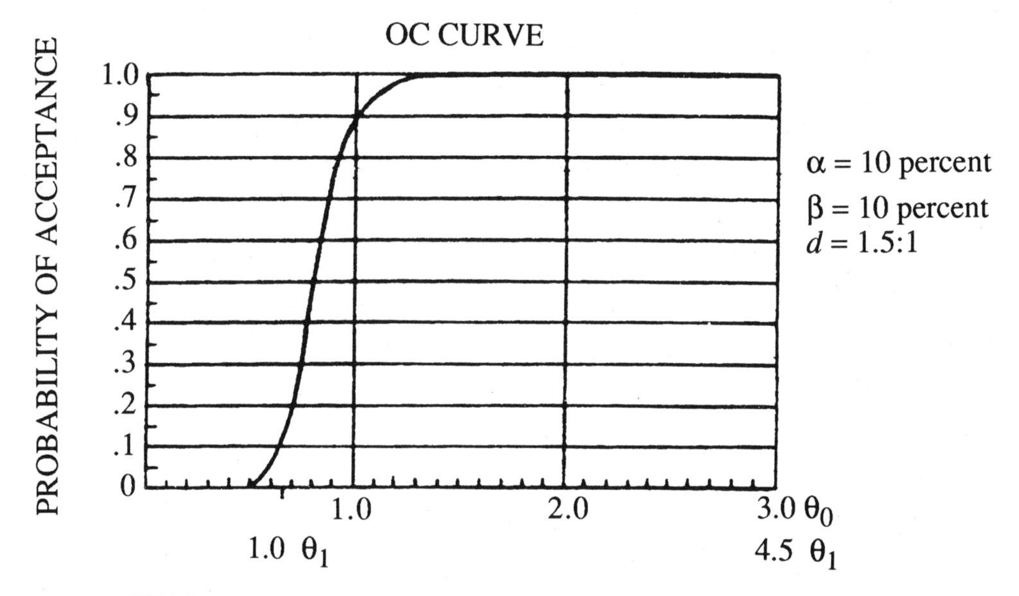

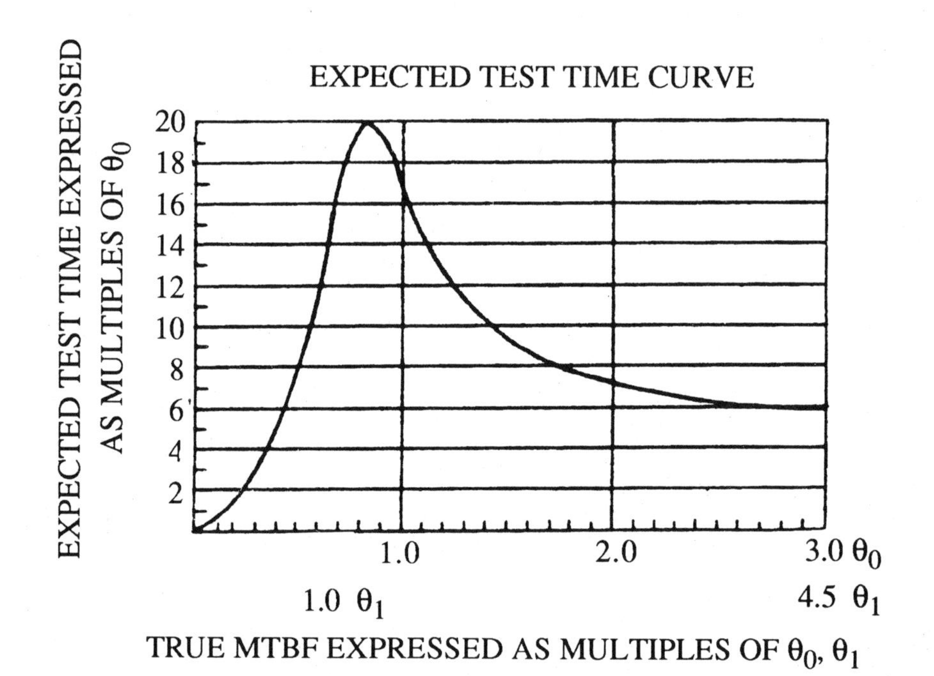

FIGURE A.3(a) · *Test Plan I-D (MIL-HDBK-781 Figure 9)*

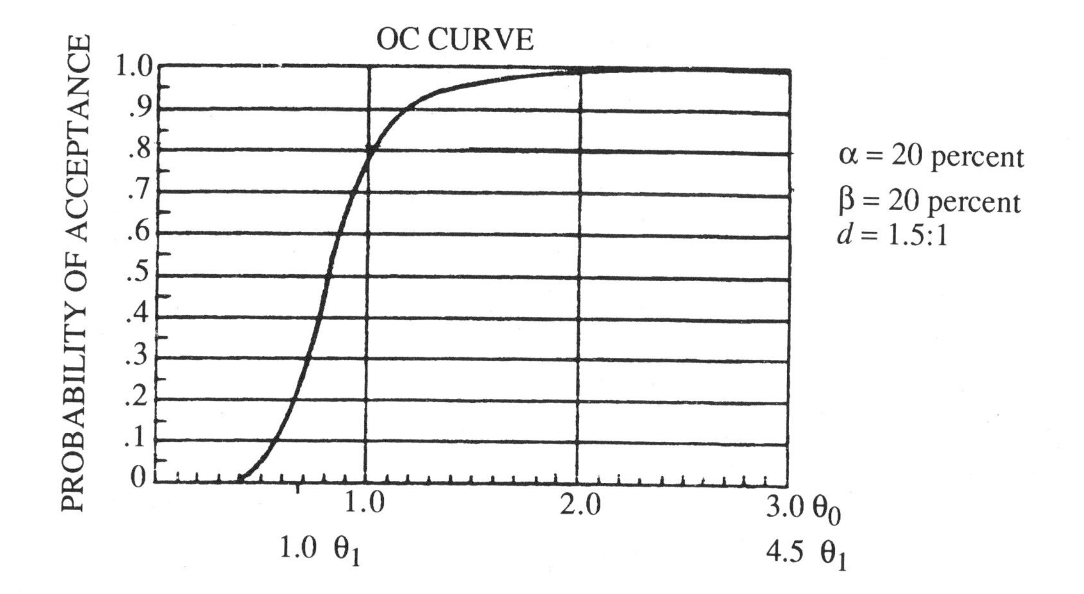

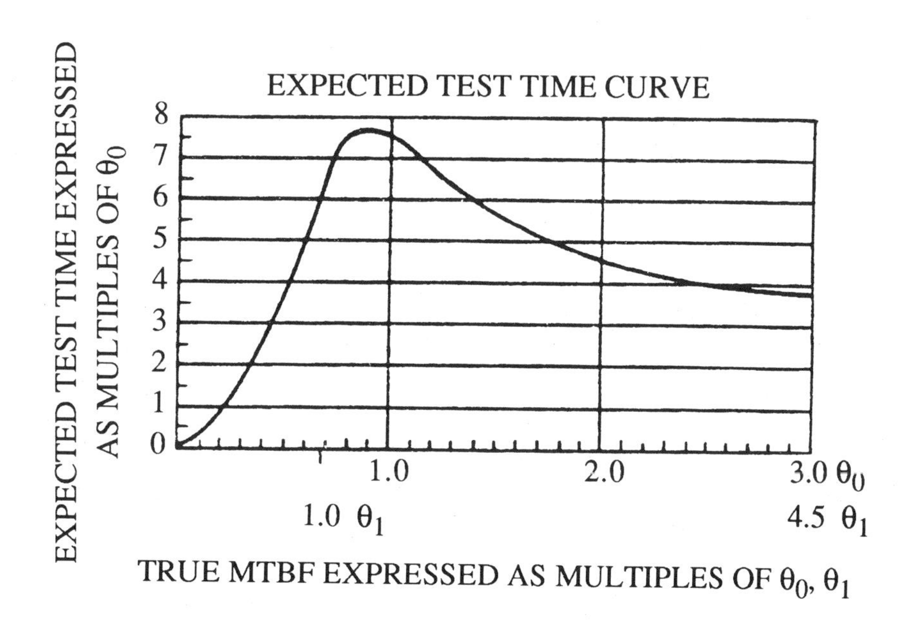

FIGURE A.3(b) · *Test Plan II-D (MIL-HDBK-781 Figure 10)*

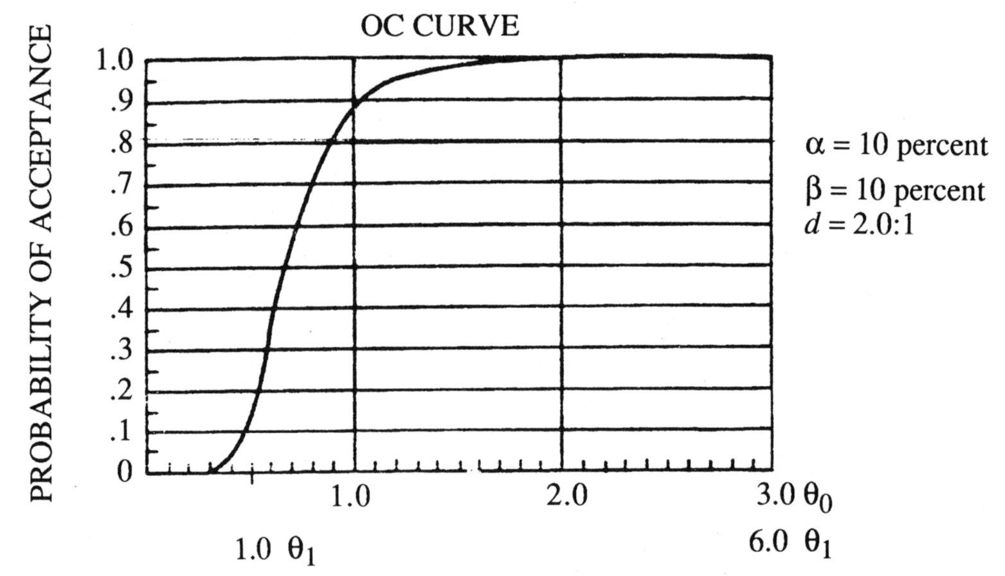

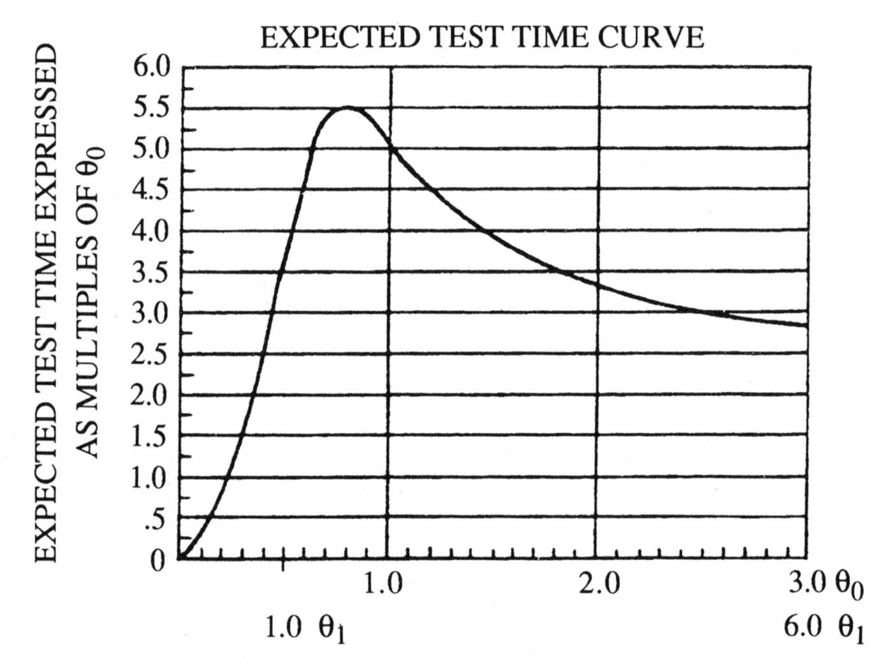

FIGURE A.3(c) · *Test Plan III-D (MIL-HDBK-781 Figure 11)*

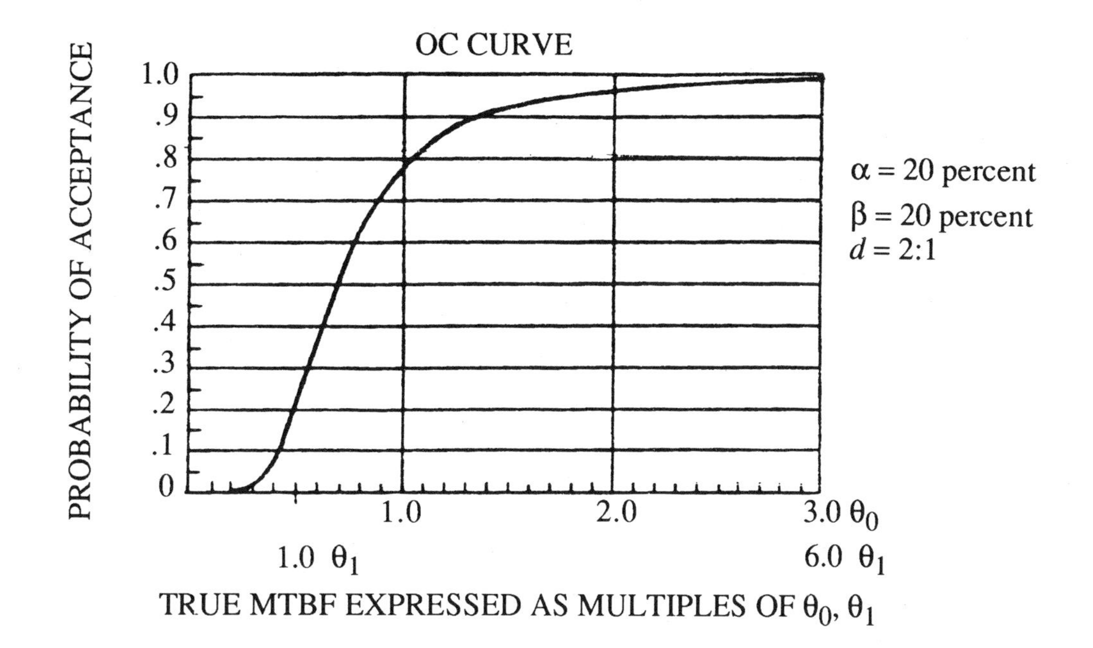

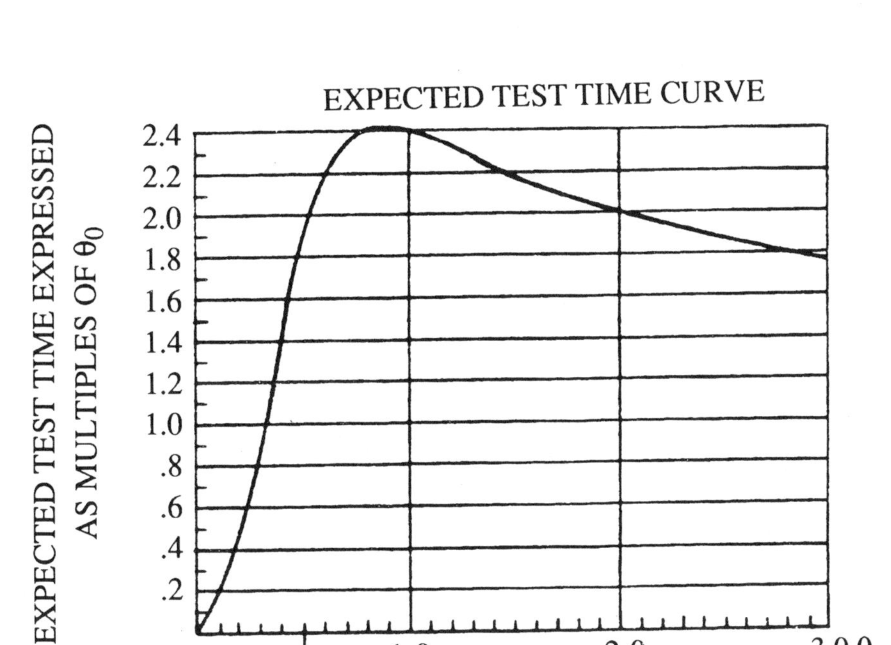

FIGURE A.3(d) · *Test Plan IV-D (MIL-HDBK-781 Figure 12)*

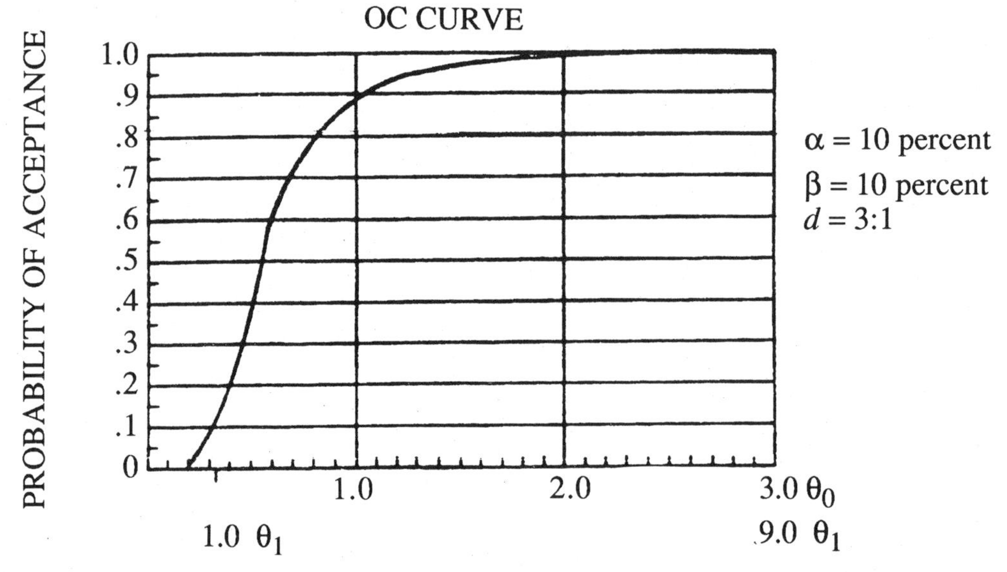

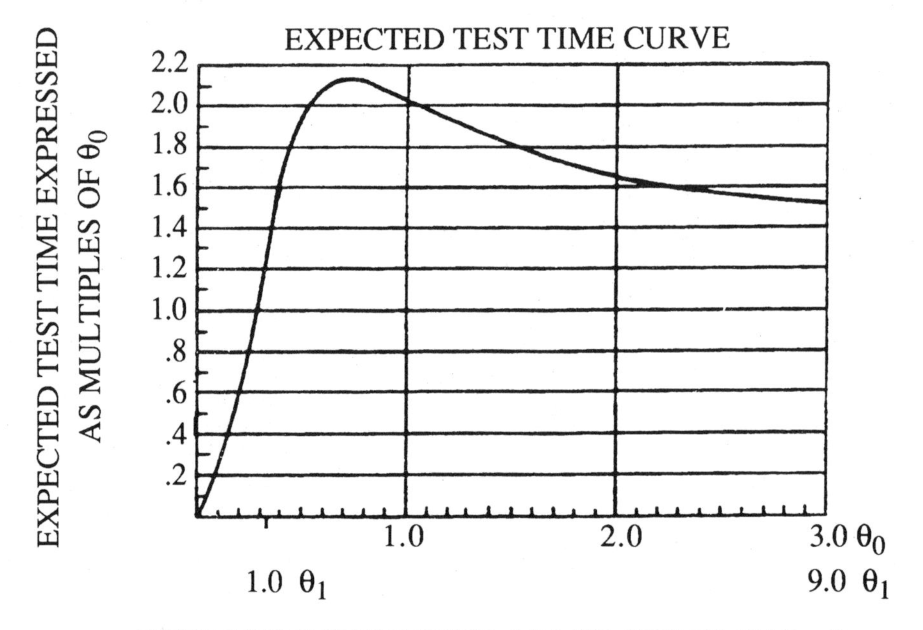

FIGURE A.3(e) · *Test Plan V-D (MIL-HDBK-781 Figure 13)*

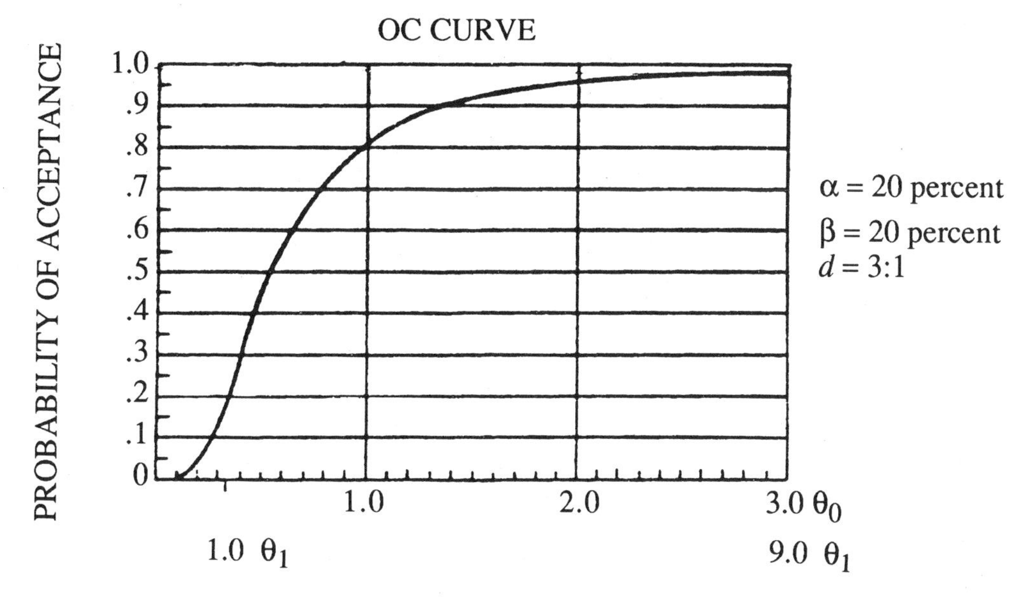

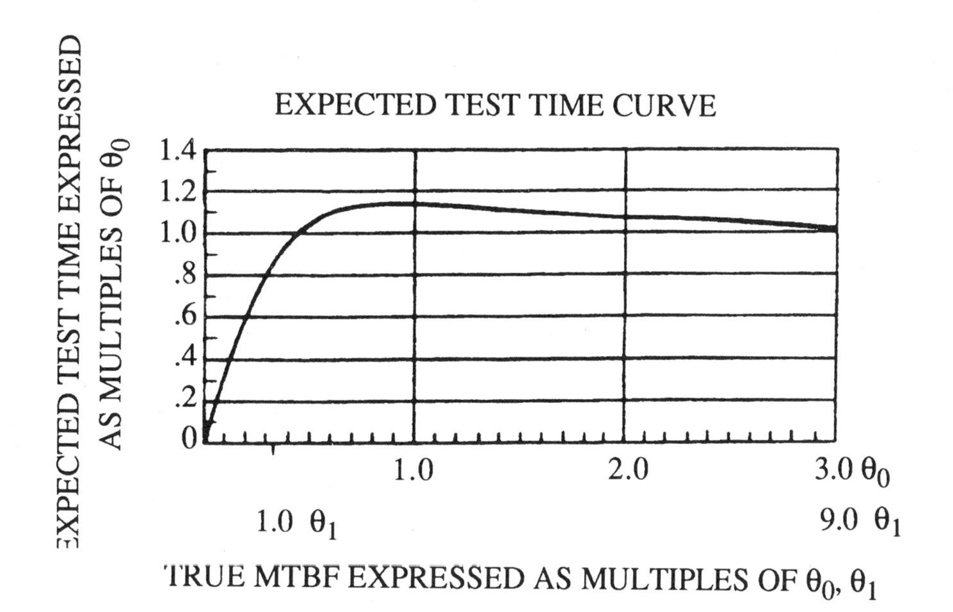

FIGURE A.3(f) · *Test Plan VI-D (MIL-HDBK-781 Figure 14)*

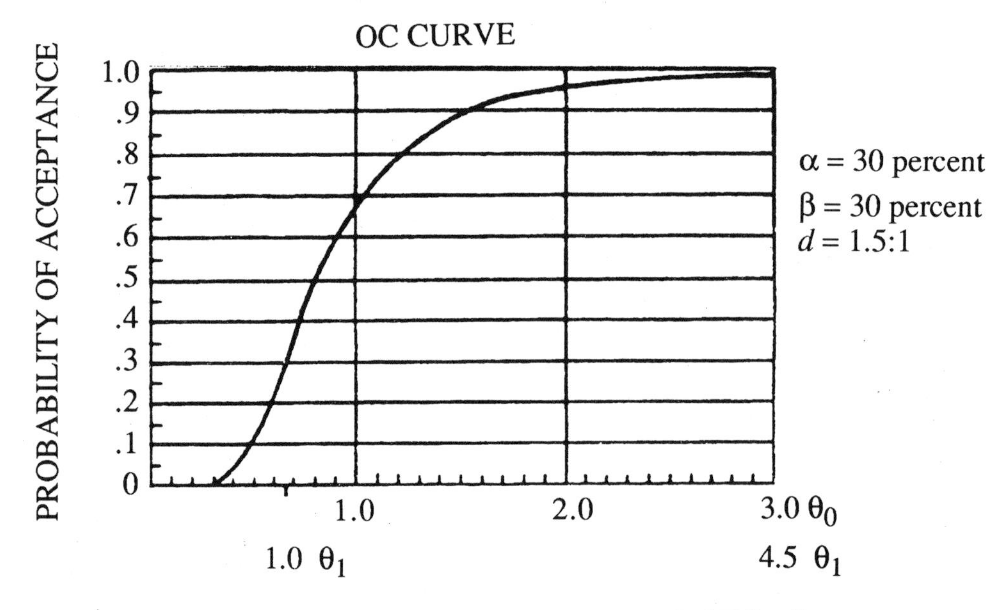

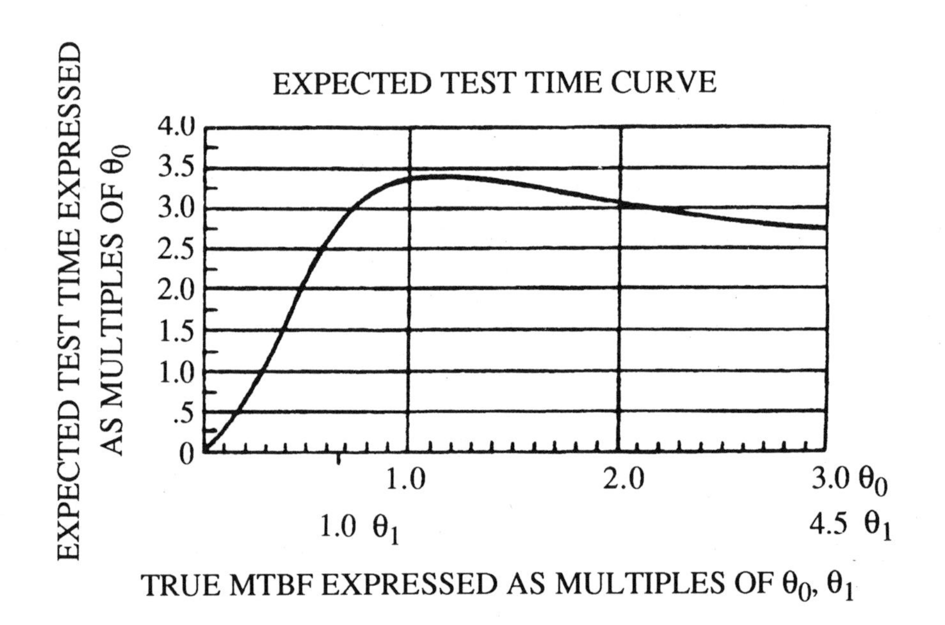

FIGURE A.3(g) · *Test Plan VII-D (MIL-HDBK-781 Figure 15)*

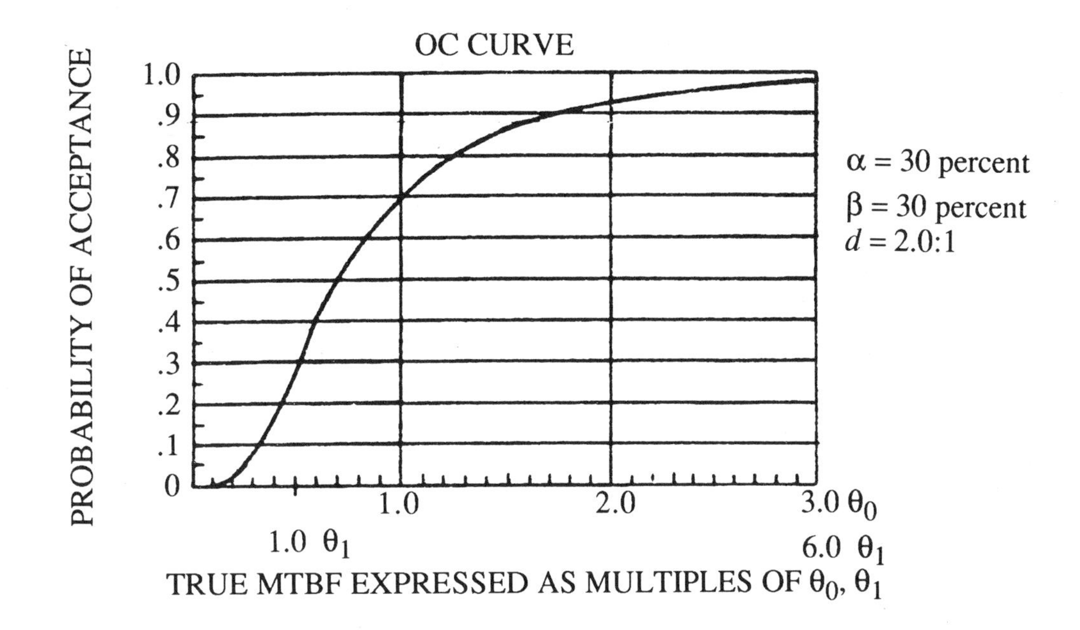

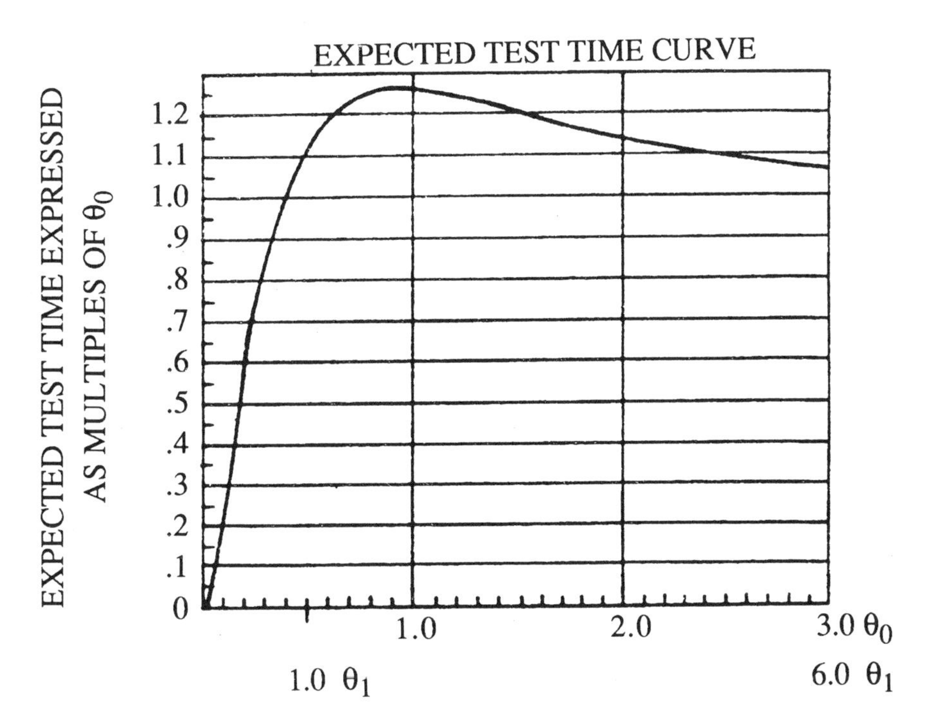

FIGURE A.3(h) · *Test Plan VIII-D (MIL-HDBK-781 Figure 16)*

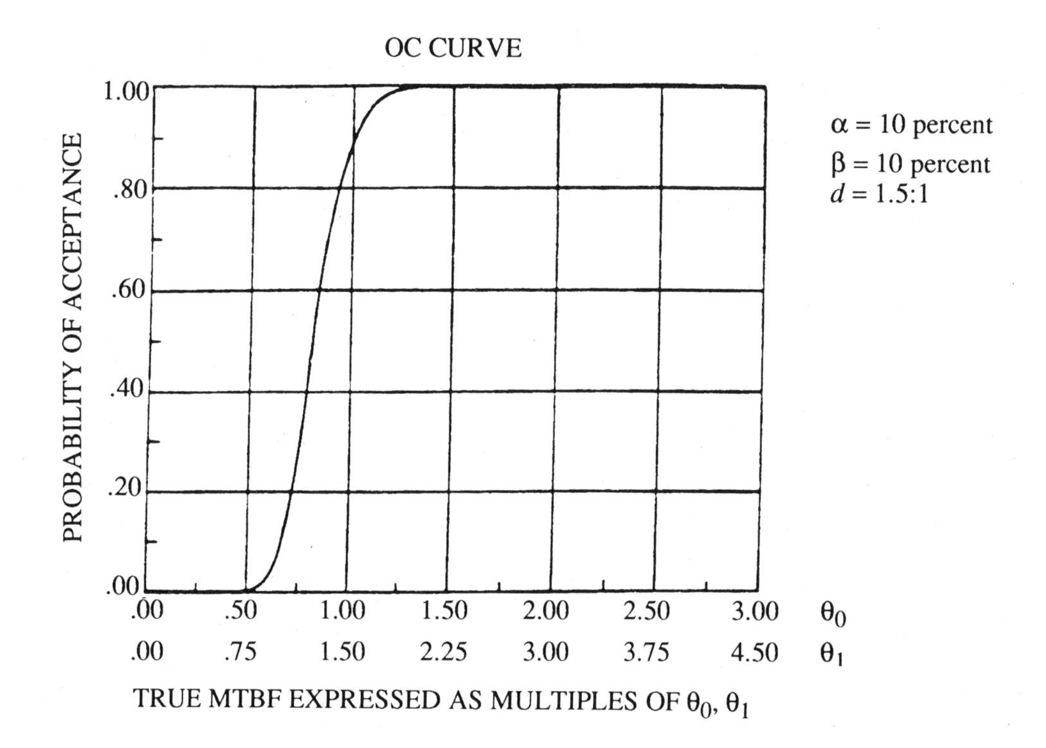

FIGURE A.4(a) · *Test Plan IX-D (MIL-HDBK-781 Figure 23)*

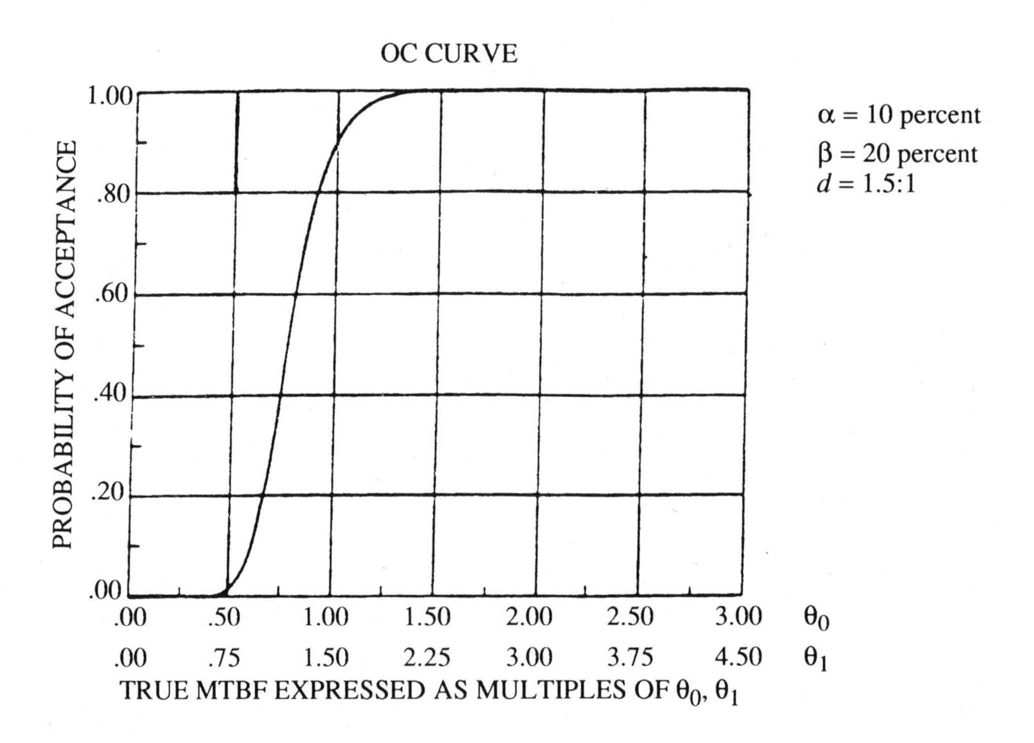

FIGURE A.4(b) · *Test Plan X-D (MIL-HDBK-781 Figure 24)*

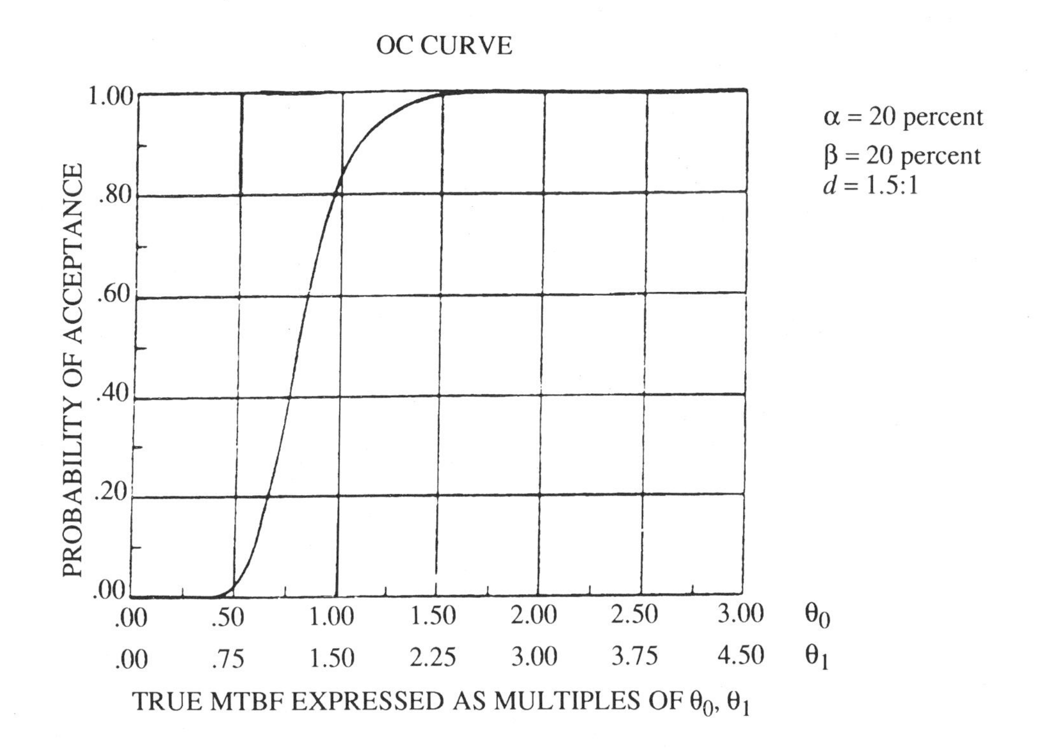

FIGURE A.4(c) · *Test Plan XI-D (MIL-HDBK-781 Figure 25)*

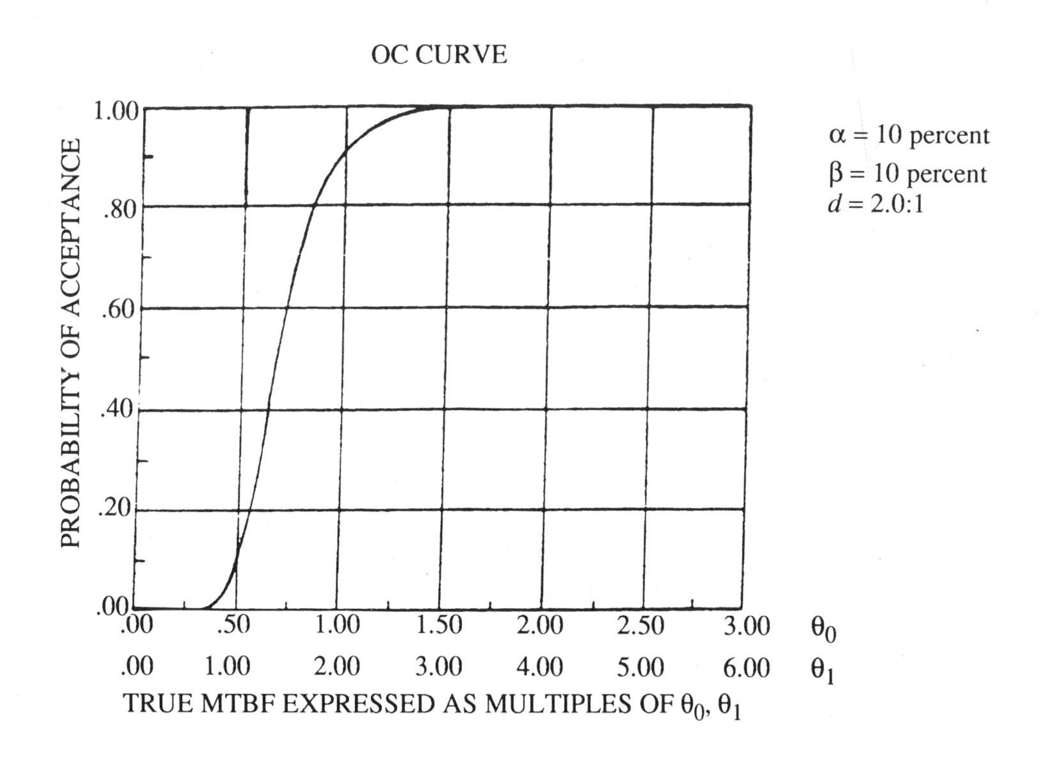

FIGURE A.4(d) · *Test Plan XII-D (MIL-HDBK-781 Figure 26)*

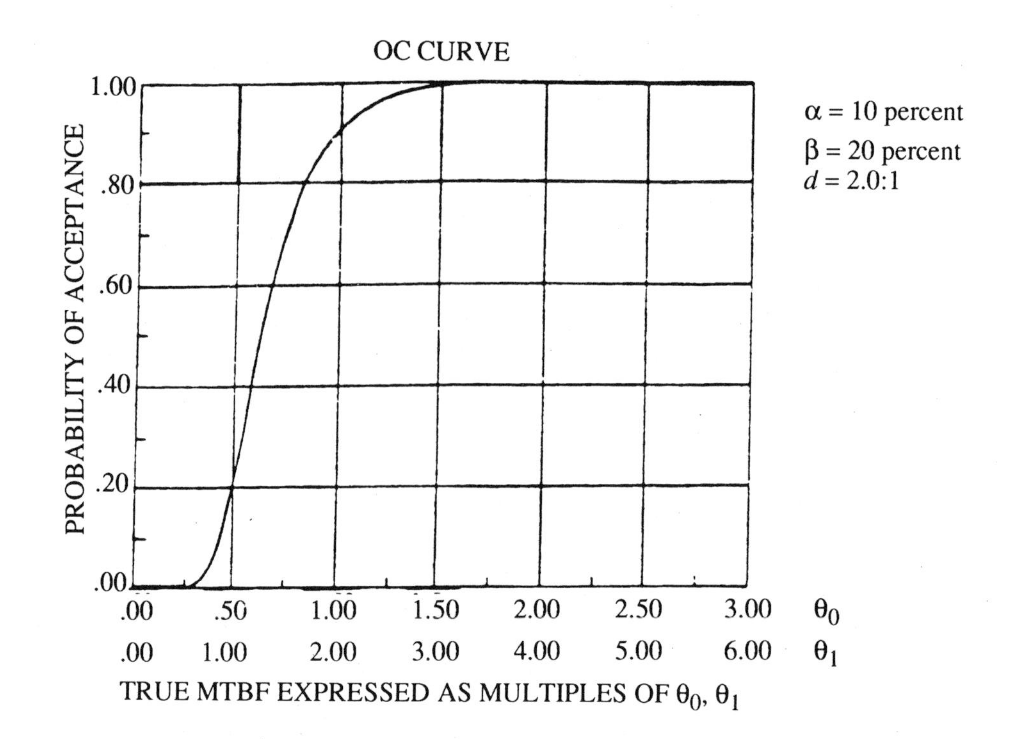

FIGURE A.4(e) · *Test Plan XIII-D (MIL-HDBK-781 Figure 27)*

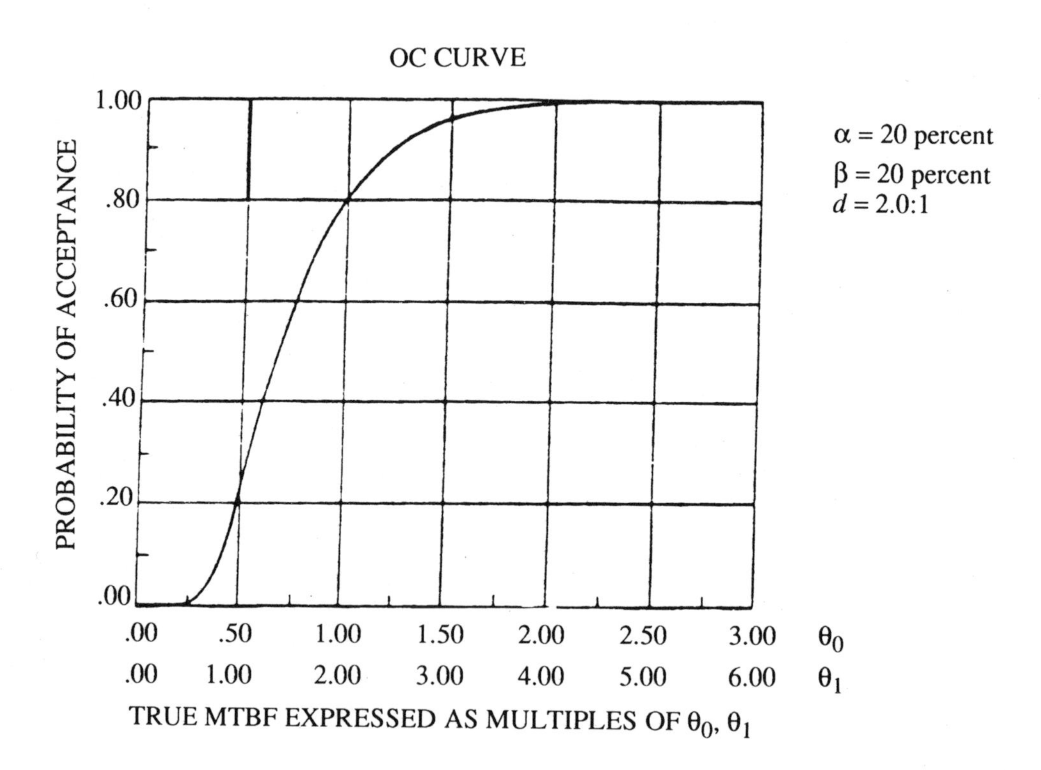

FIGURE A.4(f) · *Test Plan XIV-D (MIL-HDBK-781 Figure 28)*

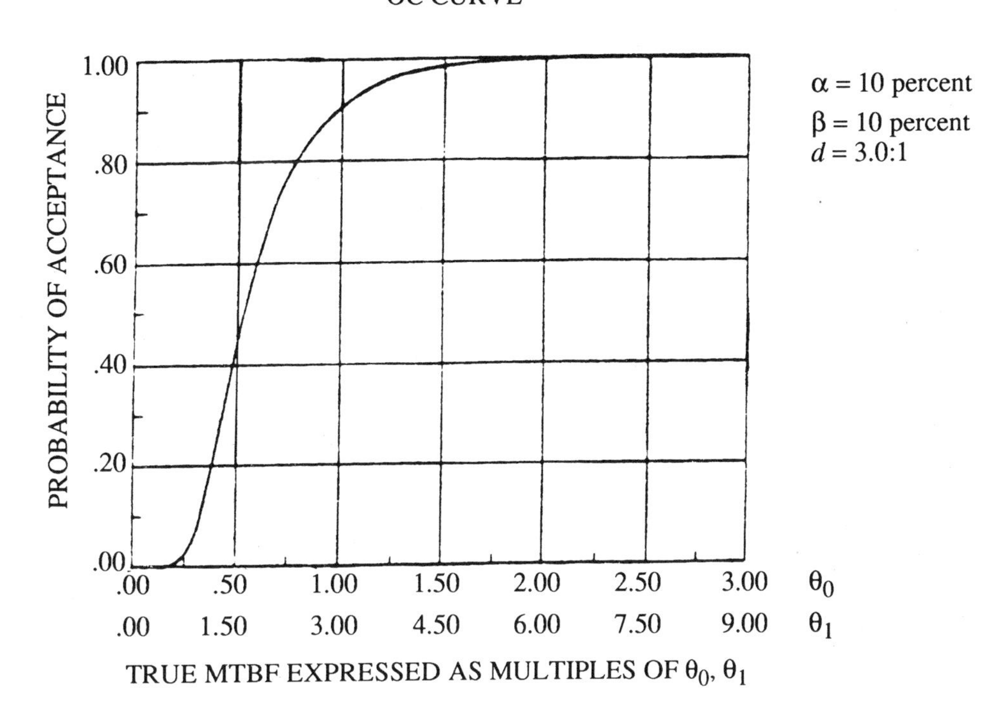

FIGURE A.4(g) · *Test Plan XV-D (MIL-HDBK-781 Figure 29)*

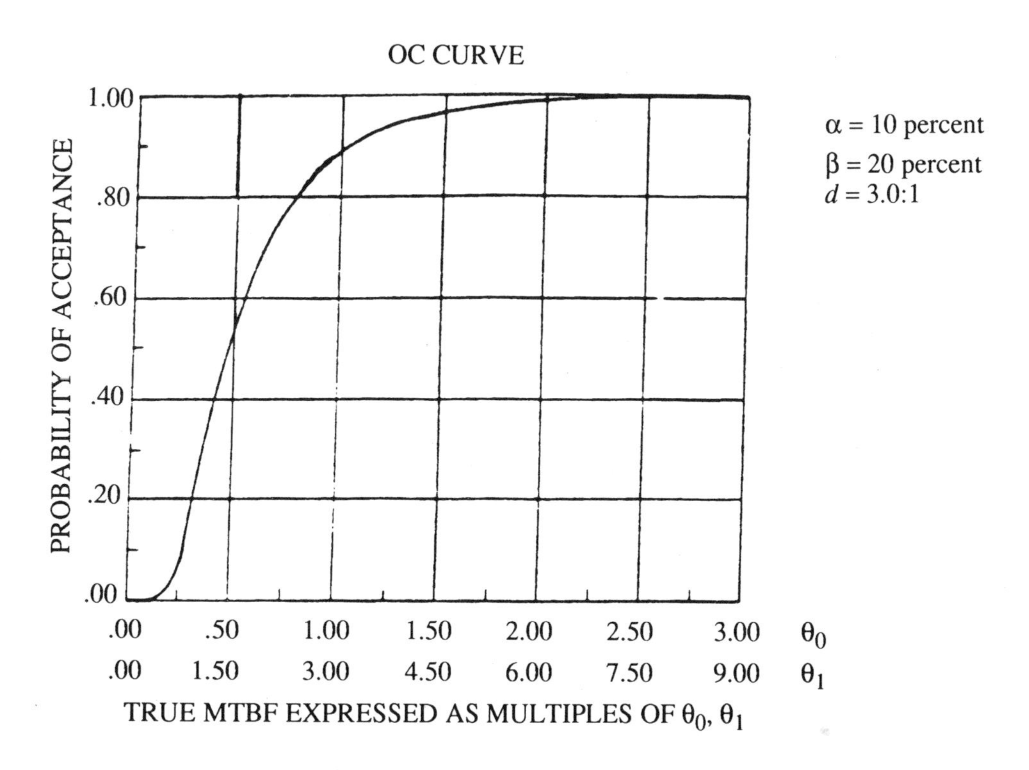

FIGURE A.4(h) · *Test Plan XVI-D (MIL-HDBK-781 Figure 30)*

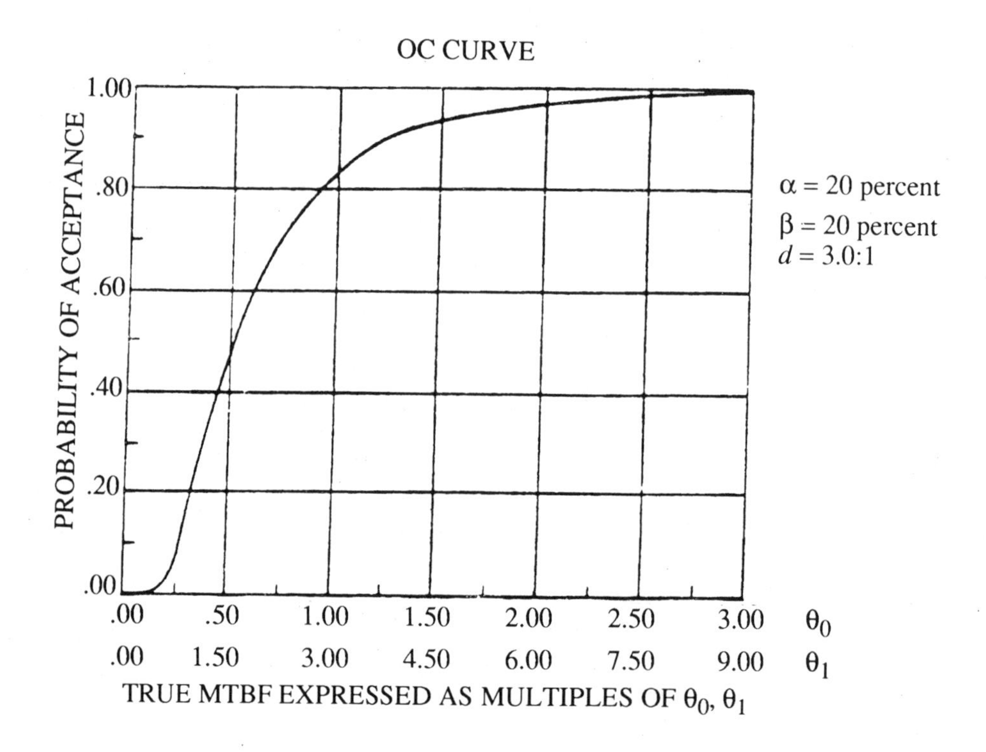

FIGURE A.4(i) · *Test Plan XVII–D (MIL–HDBK–781 Figure 31)*

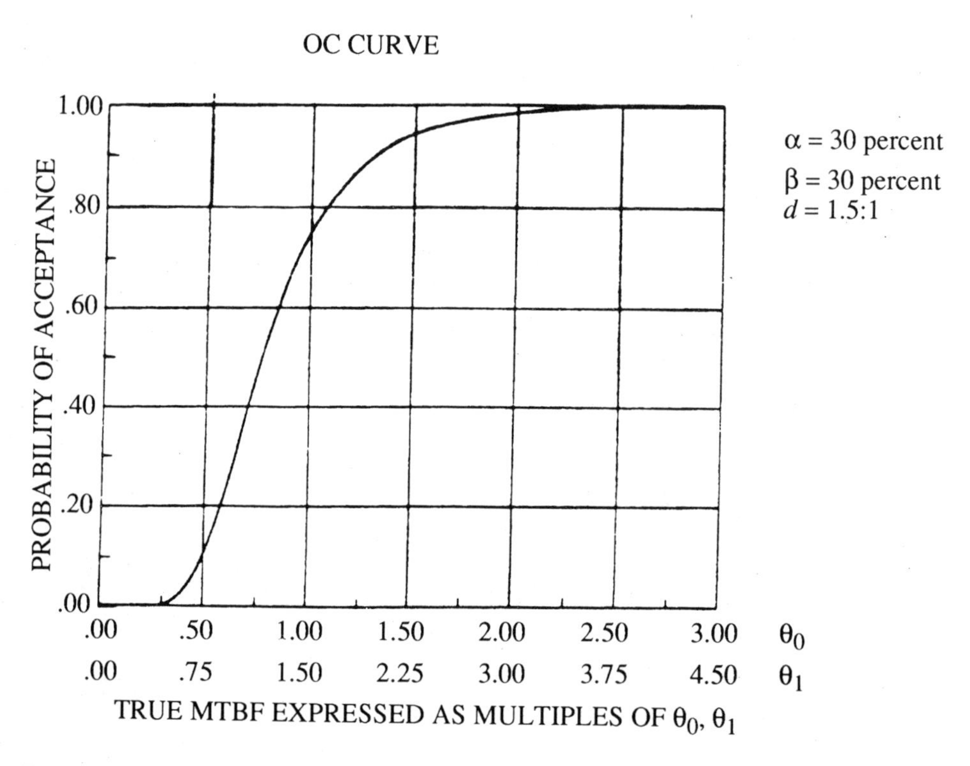

FIGURE A.4(j) · *Test Plan XIX–D (MIL–HDBK–781 Figure 32)*

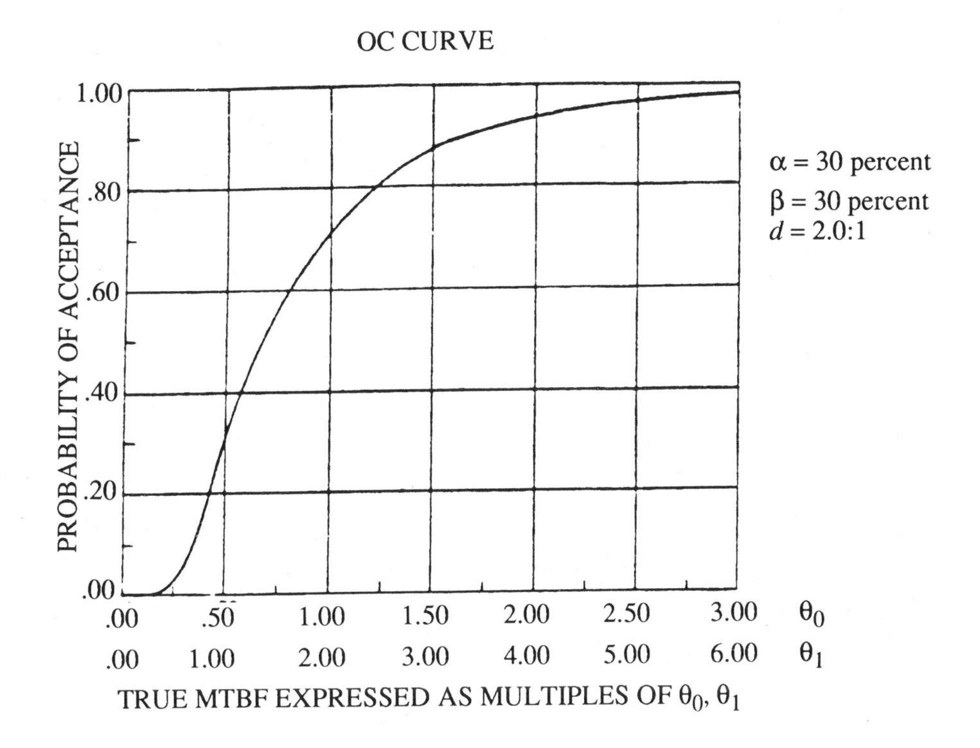

FIGURE A.4(k) · *Test Plan XX-D (MIL-HDBK-781 Figure 33)*

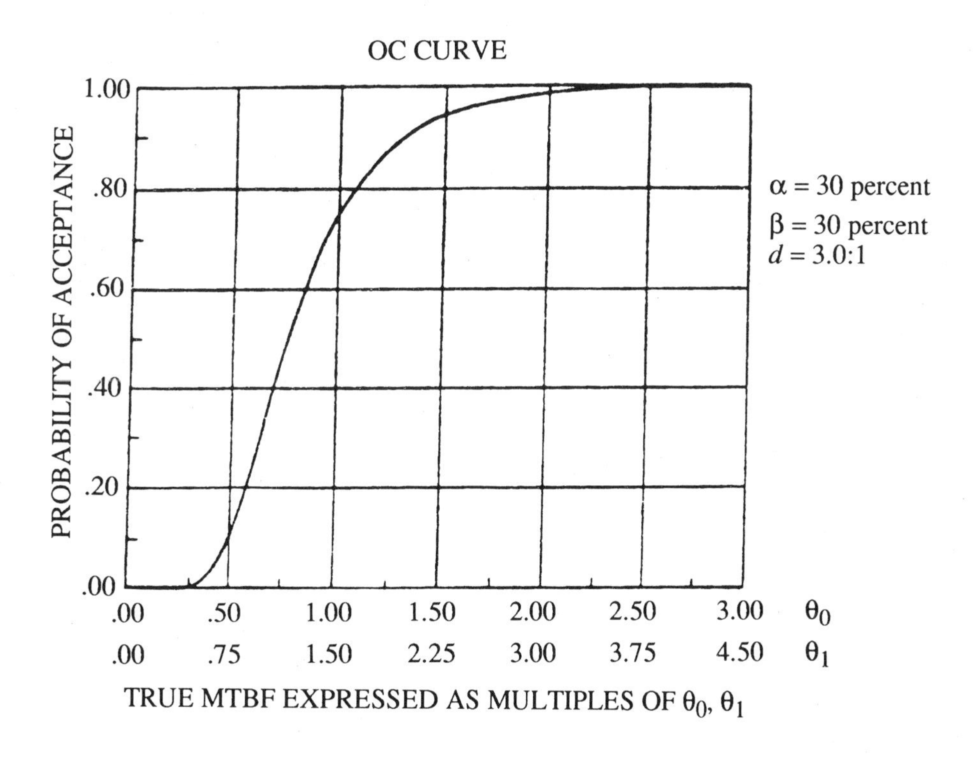

FIGURE A.4(l) · *Test Plan XXI-D (MIL-HDBK-781 Figure 34)*

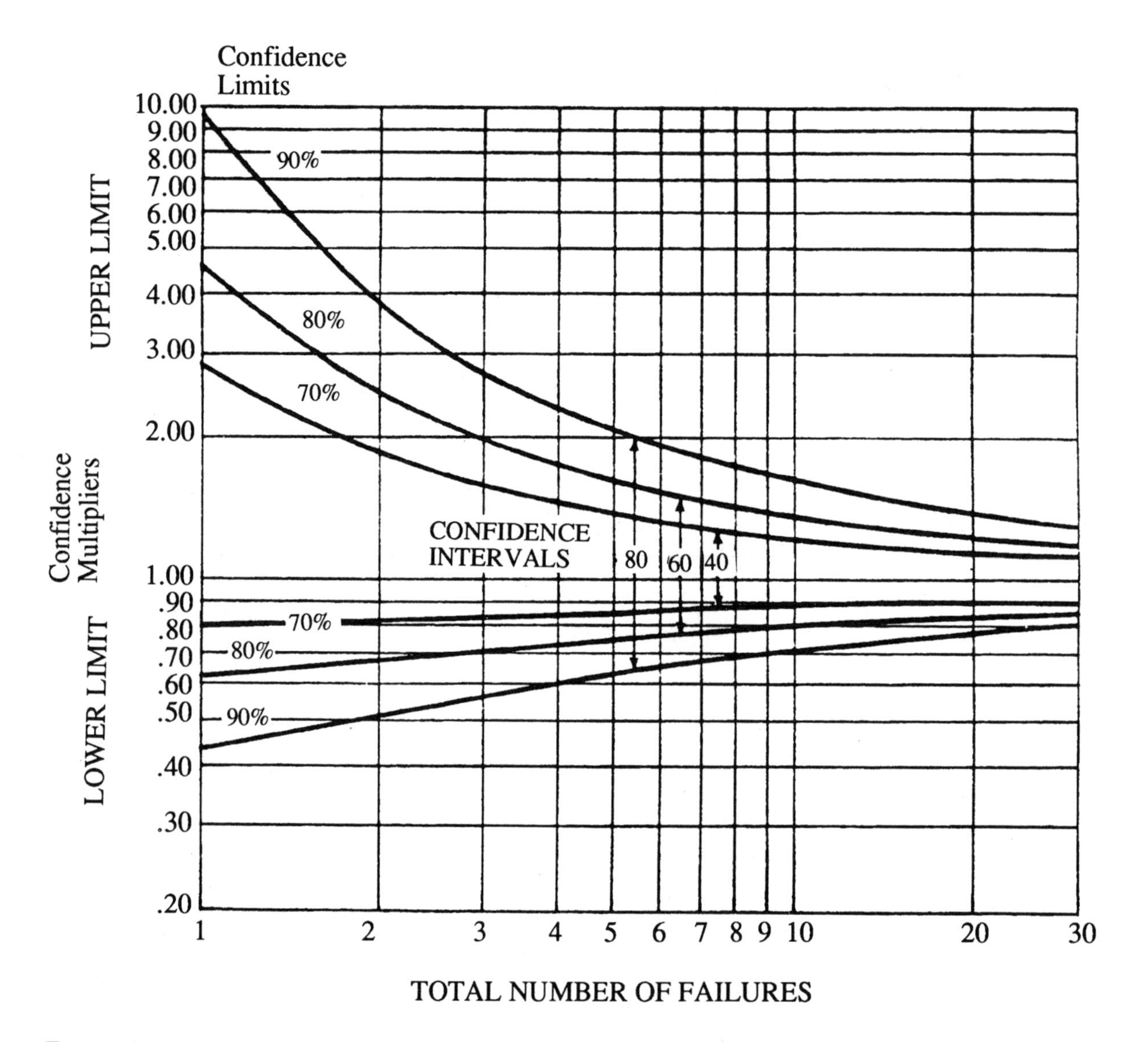

FIGURE A.5 · *Demonstrated MTBF confidence limit multipliers for failure calculation (MIL-HDBK-781 Figure 21)*

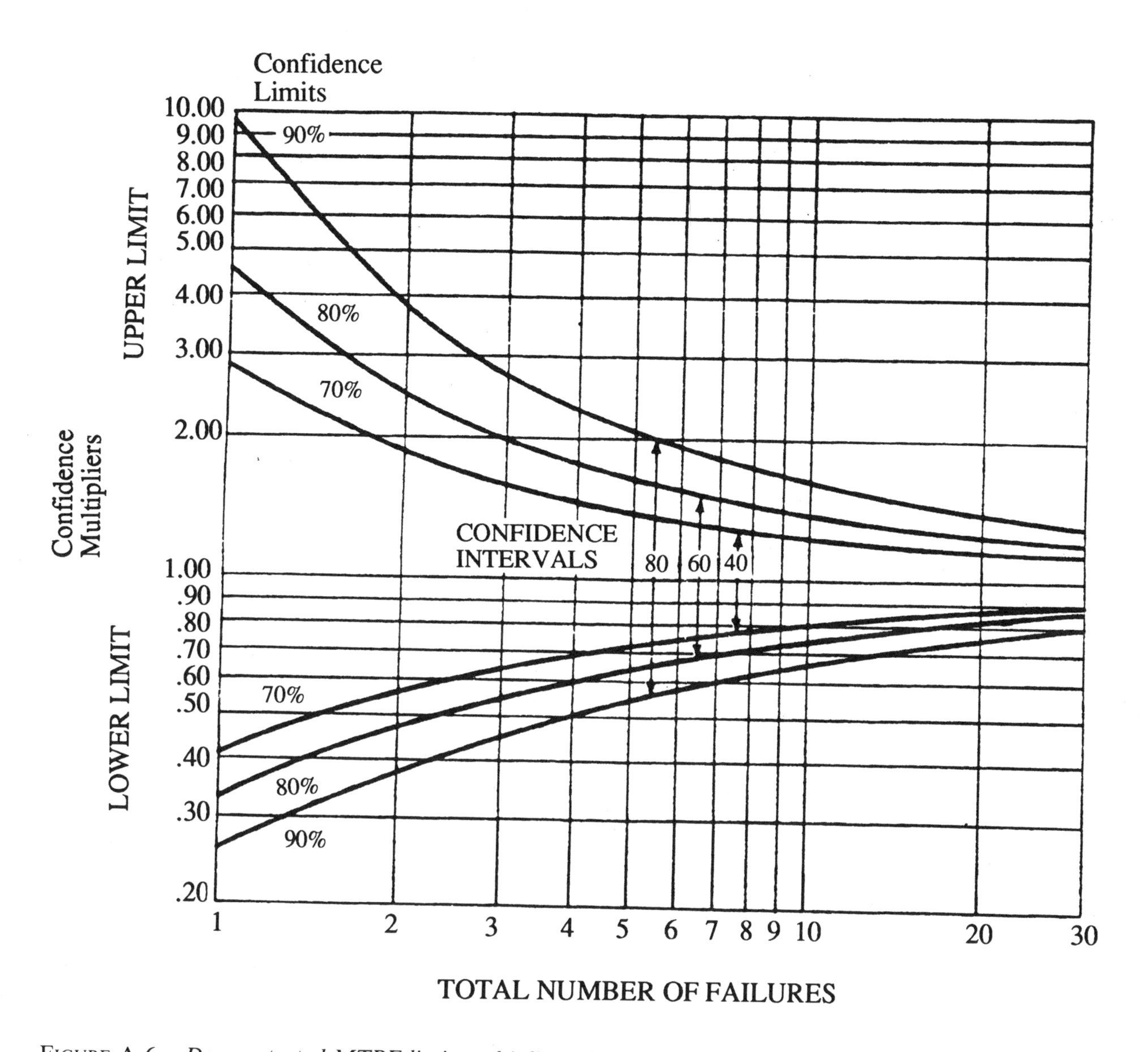

FIGURE A.6 · *Demonstrated MTBF limit multipliers, for time calculation (MIL-HDBK-781 Figure 22)*

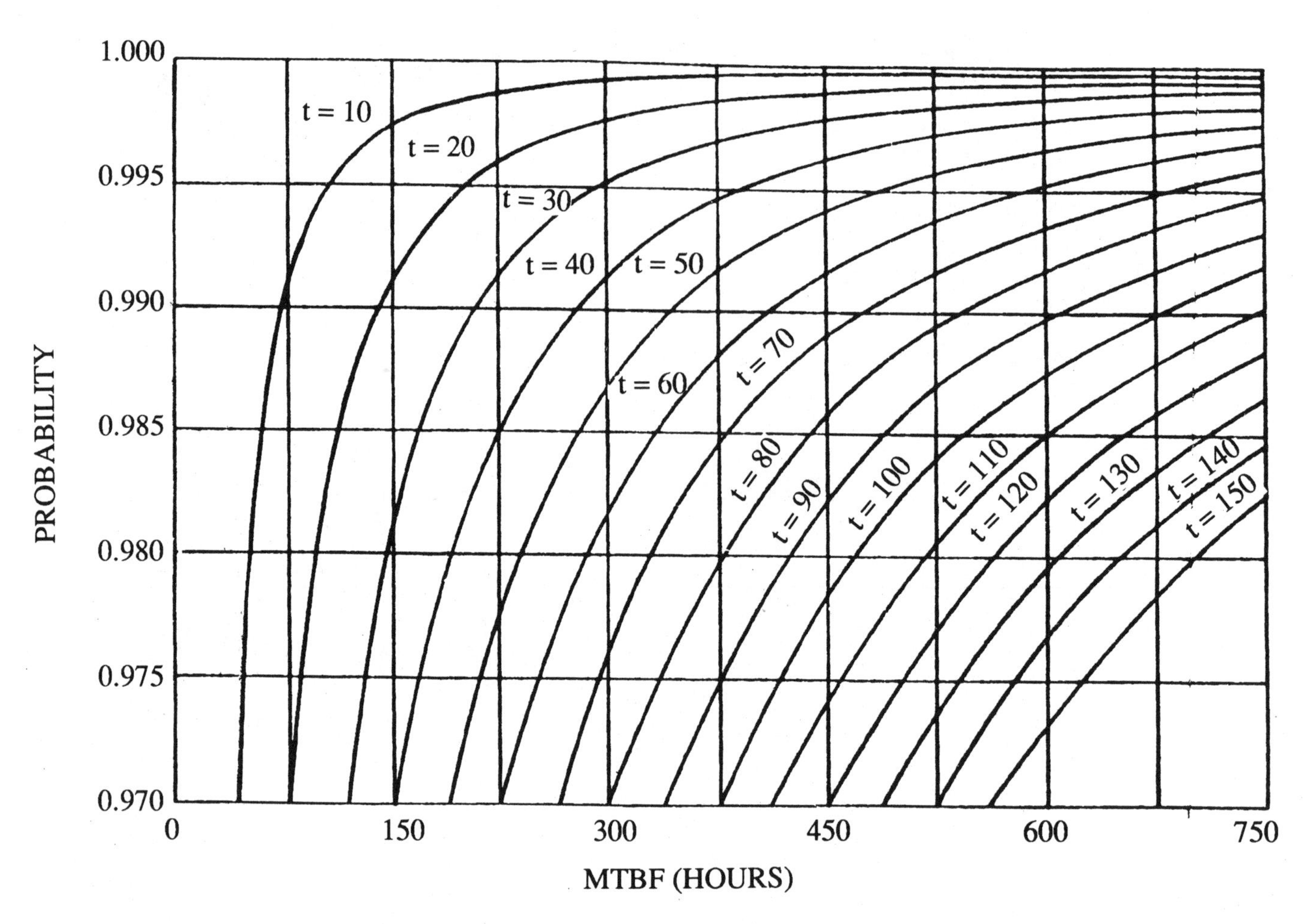

FIGURE A.7 · *MTBF assurance test (curve 1) (MIL-HDBK-781 Figure 6)*

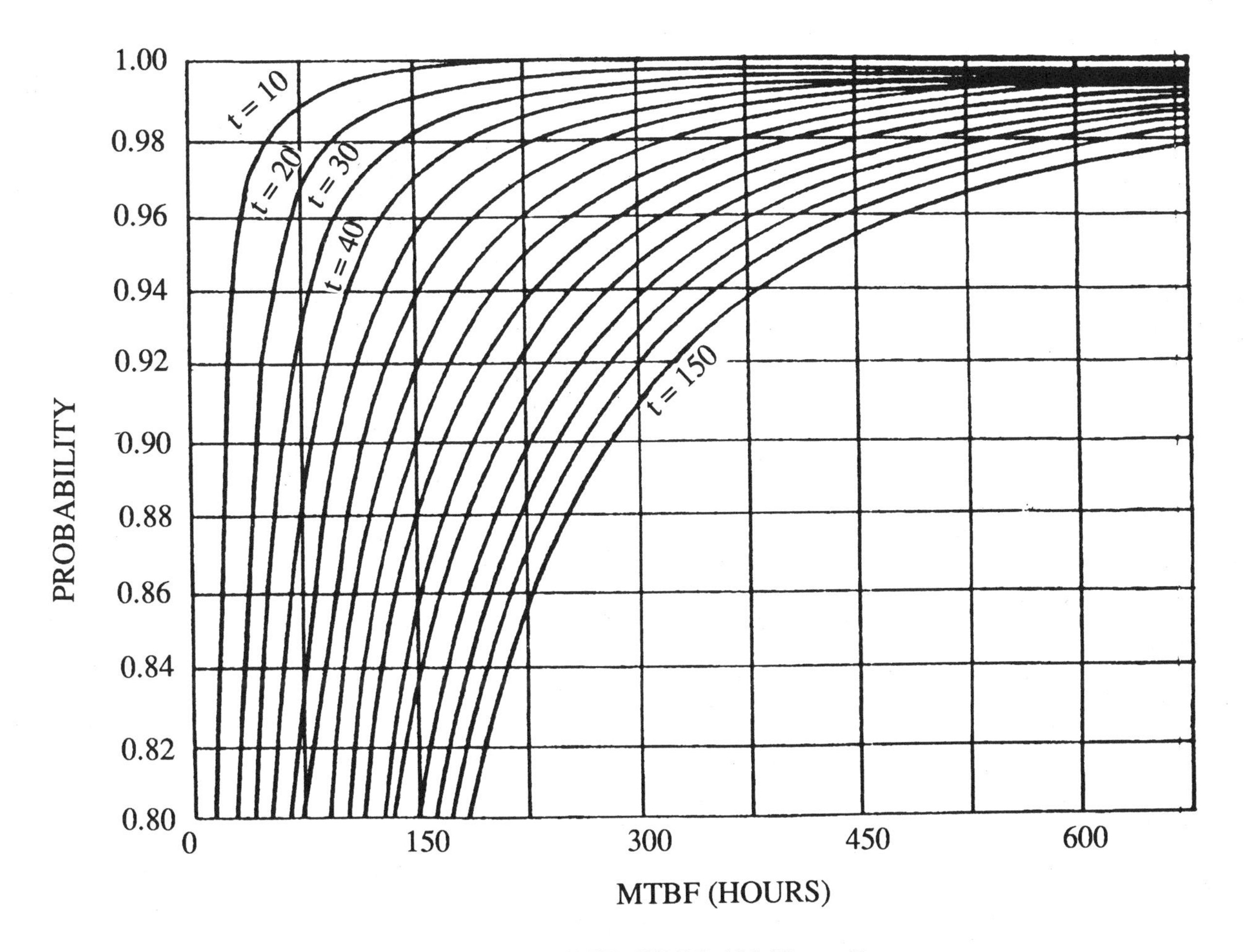

FIGURE A.8 · *MTBF assurance test (curve 2) (MIL–HDBK–781 Figure 7)*

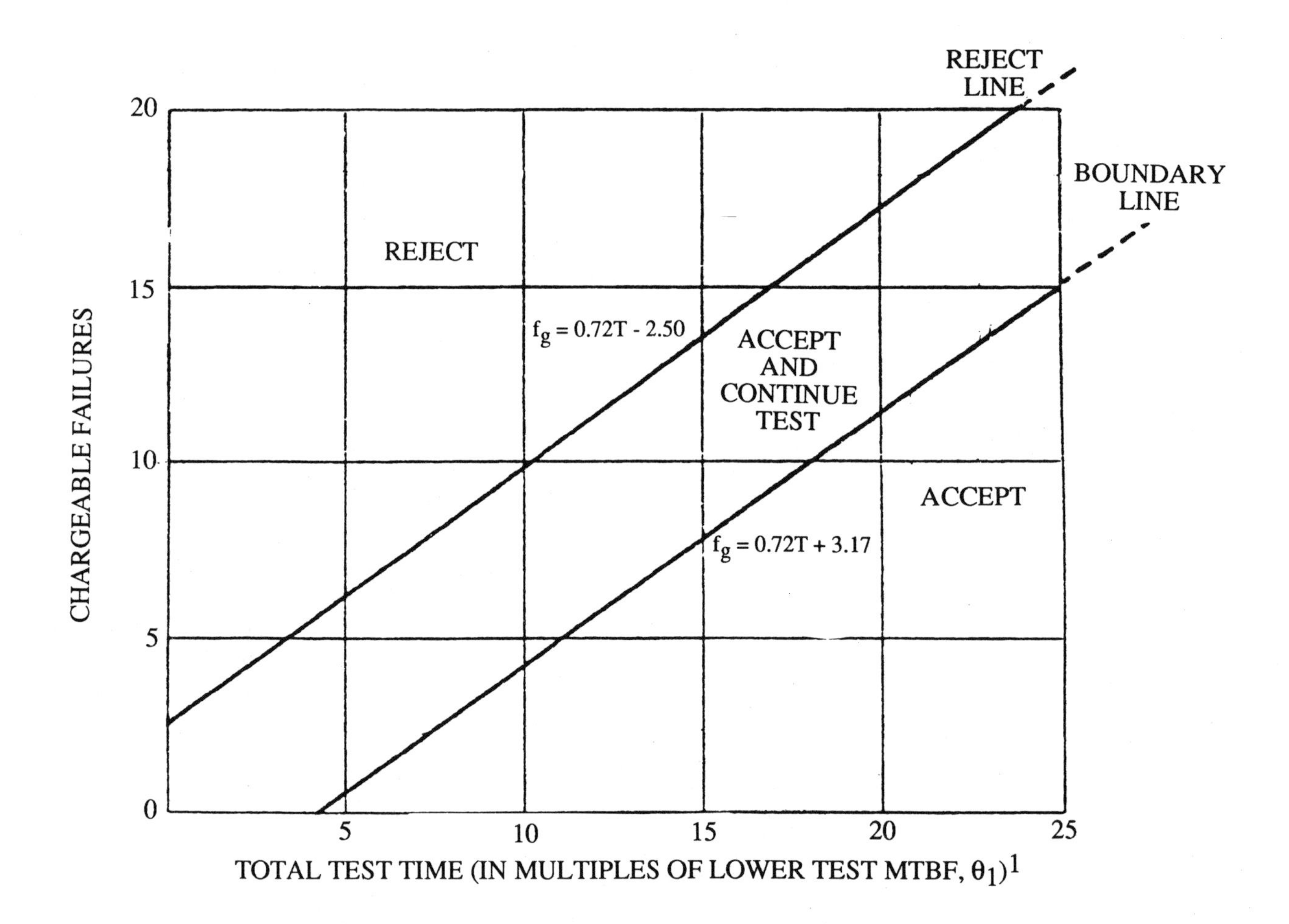

Chargeable Failures	Standardized Test Time, t^2		Chargeable Failures	Standardized Test Time, t^2	
	Reject Line	Boundary Line		Reject Line	Boundary Line
0	N/A	4.40	9	9.02	16.88
1	N/A	5.79	10	10.40	18.26
2	N/A	7.18	11	11.79	19.65
3	0.70	8.56	12	13.18	21.04
4	2.08	9.94	13	14.56	22.42
5	3.48	11.34	14	15.95	23.81
6	4.86	12.72	15	17.33	25.19
7	6.24	14.10	16	18.72	26.58
8	7.63	15.49			

Accept-reject criteria

[1]Total test time is the summation of operating time of all units included in test sample.

[2]To determine the actual test time, multiply the standardized test time (t) by the lower test MTBF (Θ_1).

ALL-EQUIPMENT PRODUCTION RELIABILITY ACCEPTANCE TEST PLAN.

FIGURE A.9 · *Test Plan XVIII-D (MIL-HDBK-781 Figure 35)*

INDEX

ARUN.S.M...